A Review of Decontamination and Decommissioning Technology Development Programs at the Department of Energy

Committee on Decontamination and Decommissioning

Board on Radioactive Waste Management
Commission on Geosciences, Environment, and Resources
National Research Council

NATIONAL ACADEMY PRESS
Washington, D.C. 1998

NOTICE: The project that is the subject of this report was approved by the Governing Board of the National Research Council, whose members are drawn from the councils of the National Academy of Sciences, the National Academy of Engineering, and the Institute of Medicine. The members of the committee responsible for the report were chosen for their special competences and with regard for appropriate balance.

Support for this study was provided by the U.S. Department of Energy under Grant No. DE-FC01-94EW54069. All opinions, findings, conclusions, or recommendations expressed herein are those of the authors and do not necessarily reflect the views of the U.S. Department of Energy.

International Standard Book Number 0-309-06281-0

Additional copies of this report are available from:

National Academy Press
2101 Constitution Avenue, N.W.
Box 285
Washington, DC 20055
800-624-6242
202-334-3313 (in the Washington Metropolitan Area)
http://www.nap.edu

Cover: Demonstration of light-aided technologies for Hanford decontamination and decommissioning projects. Photo courtesy of the Department of Energy's Deactivation and Decommissioning Focus Area.

COMMITTEE ON DECONTAMINATION AND DECOMMISSIONING

PETER B. MYERS, *Chair*, Consultant, Washington, D.C.
PATRICIA ANN BAISDEN, Lawrence Livermore National Laboratory,
  Oakland, California
SOL BURSTEIN, Wisconsin Electric Power Company (retired), Milwaukee
JOSEPH S. BYRD, University of South Carolina (retired), Lexington
BRUCE CLEMENS, James Madison University, Harrisonburg, Virginia
FRANK CRIMI, Lockheed Environmental Systems (retired), Saratoga,
  California
MILTON LEVENSON, Bechtel International (retired), Menlo Park, California
RAY O. SANDBERG, Bechtel (retired), Moraga, California
ALFRED SCHNEIDER, Georgia Institute of Technology (retired), Dunwoody
LINDA WENNERBERG, Environmental Business Systems, Cambridge,
  Massachusetts

NRC Staff

JOHN R. WILEY, Senior Staff Officer (January 1998 to present)
KARYANIL THOMAS, Senior Staff Officer (beginning of project to
  December 1997)
ROBIN L. ALLEN, Senior Project Assistant
TONI GREENLEAF, Senior Project Assistant (1997)

iv

The National Academy of Sciences is a private, nonprofit, self-perpetuating society of distinguished scholars engaged in scientific and engineering research, dedicated to the furtherance of science and technology and to their use for the general welfare. Upon the authority of the charter granted to it by the Congress in 1863, the Academy has a mandate that requires it to advise the federal government on scientific and technical matters. Dr. Bruce Alberts is president of the National Academy of Sciences.

The National Academy of Engineering was established in 1964, under the charter of the National Academy of Sciences, as a parallel organization of outstanding engineers. It is autonomous in its administration and in the selection of its members, sharing with the National Academy of Sciences the responsibility for advising the federal government. The National Academy of Engineering also sponsors engineering programs aimed at meeting national needs, encourages education and research, and recognizes the superior achievements of engineers. Dr. William A. Wulf is president of the National Academy of Engineering.

The Institute of Medicine was established in 1970 by the National Academy of Sciences to secure the services of eminent members of appropriate professions in the examination of policy matters pertaining to the health of the public. The Institute acts under the responsibility given to the National Academy of Sciences by its congressional charter to be an adviser to the federal government and, upon its own initiative, to identify issues of medical care, research, and education. Dr. Kenneth I. Shine is president of the Institute of Medicine.

The National Research Council was organized by the National Academy of Sciences in 1916 to associate the broad community of science and technology with the Academy's purposes of furthering knowledge and of advising the federal government. Functioning in accordance with general policies determined by the Academy, the Council has become the principal operating agency of both the National Academy of Sciences and the National Academy of Engineering in providing services to the government, the public, and the scientific and engineering communities. The Council is administered jointly by both Academies and the Institute of Medicine. Dr. Bruce Alberts and Dr. William A. Wulf are the chairman and vice-chairman, respectively, of the National Research Council.

Preface

This report has been reviewed in draft form by individuals chosen for their diverse perspectives and technical expertise, in accordance with procedures approved by the NRC's Report Review Committee. The purpose of this independent review is to provide candid and critical comments that will assist the institution in making the published report as sound as possible and to ensure that the report meets institutional standards for objectivity, evidence, and responsiveness to the study charge. The review comments and draft manuscript remain confidential to protect the integrity of the deliberative process. We wish to thank the following individuals for their participation in the review of this report:

Harold Agnew, Los Alamos National Laboratory & General Atomics (retired)
Edward Albenesius, Savannah River Site (retired)
Robert Budnitz, Future Resources Associates, Inc.
Gerald Dinneen, Honeywell, Inc. (retired)
James Johnson, Jr., Howard University
Richard Meserve, Covington & Burling
George Parshall, DuPont (retired)
William Prindle, Corning (retired)
Dennis Reisenweaver, International Atomic Energy Agency
Richard Smith, Pacific Northwest National Laboratory (retired)
George Whitesides, Harvard University
Raymond Wymer, Oak Ridge National Laboratory (retired)

While the individuals listed above have provided constructive comments and suggestions, it must be emphasized that responsibility for the final content of this report rests entirely with the authoring committee and the institution.

Contents

APPENDIXES

Executive Summary

More than 7,000 radioactively contaminated buildings located on U.S. Department of Energy (DOE) sites that formerly produced nuclear defense materials must be decontaminated, and about 700 buildings are to be decommissioned fully. In addition, concrete and some 180,000 metric tons of scrap metal within these facilities must be dealt with. This decontamination and decommissioning (D&D) work is a significant part of the overall DOE Office of Environmental Management (EM) program to remediate the Cold War legacy.

The Office of Science and Technology (OST) within EM (EM-50) is charged with assuring that safe, cost-effective technologies are available for the entire remediation program. To address D&D technology needs, OST established and staffed a program, now referred to as the Deactivation and Decommissioning Focus Area (DDFA), in 1994.[1] At the request of OST the National Research Council (NRC) convened a Committee on Decontamination and Decommissioning to assess the utility and effectiveness of programmatic approaches used by the DDFA during the period 1996-97.[2] An initial review of all of OST's technology development activities was completed in 1995 (NRC, 1996).

---

[1]The DDFA was one of five focus areas established in 1994. A brief history of EM-50 and its programs is given in Chapter 1.

[2]To be referred to as the D&D committee (or the committee) throughout this report. This committee is successor to a subcommittee (of the same name) of the Committee on Environmental Management Technologies (CEMT). The CEMT subcommittees were reorganized as independent committees under the National Research Council's Board on Radioactive Waste Management in 1997. The D&D committee's Statement of Task is given in Appendix A.

The D&D committee has spent considerable time and effort reviewing both the technology and process issues that have confronted OST and its DDFA over the past two years as the DDFA has sought to identify and promote deployment of improved D&D technologies for the cleanup of DOE sites. During this time, a cornerstone of the DDFA effort has been the Large-Scale Demonstration Program (LSDP). The initial three of these demonstrations were visited and reviewed in depth by committee members. Although OST policies and programs continue to evolve, this report summarizes the committee's findings pertinent to the LSDP and the general mission of the DDFA. Regardless of any continuing changes in programs and organization, the committee believes that the lessons learned from this review will be applicable to any future D&D undertakings by EM.

SIGNIFICANT FINDINGS AND RECOMMENDATIONS

The committee found that the DDFA generally has failed to meet its objective to promote DOE site-wide deployments of new technologies. The LSDP, the main deployment approach used by the DDFA, lacked planning and did not meet its schedules or goals during the committee's review. Prompt dissemination of sufficient technical and cost data to encourage site managers to adopt successfully demonstrated technologies was not achieved, nor were the LSDPs able to overcome institutional barriers to new technologies. While the DDFA has scored some successes and made conscientious efforts to improve its technology implementation record, which the committee recognizes, less has resulted from the application of truly novel technologies than expected at the program's inception. Widespread user acceptance of the demonstrated technologies has not occurred. In his final presentation to the committee on December 10, 1997, Gerald Boyd, Acting Deputy Assistant Secretary for Science and Technology, acknowledged that the LSDP inaugurated in 1995 had not produced the acceptance and deployment of clean-up technologies that OST had hoped (Boyd, 1997b). The committee therefore recommends that OST and DDFA substantially revise the LSDP or phase it out.

Other significant findings made by the committee are the following:

1. There is a lack of top-down evaluation and prioritization of DDFA activities.[3] The committee found that existing approaches to prioritization such as the "National Needs Assessment" reflect a bottom-up approach, which is inadequate for complex-wide planning (DOE, 1996a). The Linkage Tables by which potential or available technologies were coupled with D&D needs show no discernible prioritization (DOE, 1997a). DOE/OST is aware of this situation and is in the process of updating its strategic planning (Boyd, 1997a).

---

[3]This finding was raised in the CEMT report (NRC, 1996).

2. There is no apparent linkage between the initial set of priorities given in the "National Needs Assessment" and technologies that have been selected and funded for development by the DDFA, for example, those selected for the LSDP.

3. The DDFA has no systematic plan that will encourage the deployment of new technologies. In particular, the present approaches for assessing the success of a new technology and the process by which its performance can be compared with an existing baseline technology are incomplete and inconsistent. Convincing documentation of successfully demonstrated technologies was not available during the committee's review period. DOE/OST is aware of this situation (Boyd, 1997a).

4. End states of the facilities that will undergo D&D are not defined adequately. Without defining "how clean is clean enough" the necessary technology, cost, and schedule for a D&D project cannot be determined.[4] While the definition of end states is beyond the responsibility of OST and its DDFA, it is critical to judging the success of DDFA in developing technologies that will meet EM's future needs.

5. Given that the definition of actual facility end states is beyond the responsibility of OST-DDFA, there is no barrier to their defining reasonable, target end states to serve as indicators of applications where new technologies are likely to be needed and to serve as benchmarks for evaluating new technologies. The DDFA has not made significant progress in defining such end states.

6. Because of the historical autonomy of individual DOE sites, there is no mechanism to ensure implementation of OST-developed technologies. As long as authority and responsibility continue to reside in different entities, centralized development of technologies to be deployed throughout the DOE complex will not, in the committee's view, be effective. This problem is clearly beyond OST's scope or jurisdiction, but a solution is essential for the DDFA to be successful.

7. The capabilities residing in the private, academic, and foreign sectors for providing advanced D&D technologies are not being identified or utilized by the DDFA effectively.

To assist the DDFA in improving its efficiency, especially as it revises the LSDP and makes further efforts to achieve greater deployment of new D&D technologies, the committee makes the following recommendations:

- The DDFA should improve its strategic planning.
- Top management in OST, rather than reporting elements, should evaluate and prioritize the technology needs of the operating sites.

---

[4]The many facilities in the DOE complex will be remediated to varying degrees. Some facilities will be cleaned sufficiently for re-use; others will be restored to greenfield. A plethora of possible end states therefore exists. The concept of end states is discussed in Chapter 2.

- OST and the DDFA should link all actions and funding to the prioritized needs.
- DDFA should define reasonable target end states for D&D technologies to achieve.
- DDFA should improve its approach for introducing and gaining acceptance of demonstrated technology.
- OST and DDFA should develop and apply a uniform and consistent approach to comparative technology assessment across all projects.
- DDFA should be more aware of technologies developed in the private, academic, and foreign sectors.
- DDFA should communicate its program results in a more effective and timely manner.
- DDFA should establish a better connection between university and industry programs and prioritized long-term needs.

1

Introduction and Background

This report is the result of a study requested by the U.S. Department of Energy's (DOE) Office of Science and Technology (OST) as part of its efforts to improve the effectiveness of its Deactivation and Decontamination Focus Area (DDFA). The statement of task for the study, which is provided in Appendix A, directed the National Research Council (NRC) to convene a committee to review the utility and effectiveness of key aspects of the DDFA's program:

1. Approaches used to select and evaluate alternative decontamination and decommissioning technologies;

2. Methods of seeking deployment of the selected technologies among the DOE operating sites, with emphasis on the Large-Scale Demonstration Program (LSDP);

3. Processes used to compare the advantages of the alternative technologies to currently used technologies.

In conformance with the statement of task, the committee examined the technology development activities of the DDFA. Most information was obtained from presentations made by representatives of DOE's Office of Environmental Management (EM), OST, DDFA, and its contractors, and from publicly available reports. Committee members visited three LSDP sites, interviewed site and contractor personnel, and recorded their findings (see Appendixes C-F).

In preparing this report, the committee has described the development of the DDFA within DOE and the main activities of the DDFA in the following section of this chapter. Chapter 2 provides a discussion of the findings and conclusions developed by the committee during its study. Recommendations resulting from the study are given in Chapter 3.

PROGRAM DEVELOPMENTS IN THE DEPARTMENT OF ENERGY'S OFFICE OF ENVIRONMENTAL MANAGEMENT AND IN THE DEACTIVATION AND DECOMMISSIONING FOCUS AREA

In November 1989, DOE established its Office of Environmental Management.[5] The mission of the EM program is to bring DOE sites into compliance with all applicable regulations while minimizing risks to the environment and human health and safety posed by the generation, handling, treatment, storage, transportation, and disposal of DOE waste. The undertaking was projected to cost billions of dollars each year for many decades to come and to require additional remediation technologies (DOE, 1995a). An Office of Technology Development (OTD, EM-50)[6] was formed in 1989 within EM and charged with carrying out an aggressive national program of applied research and development (technology development program) to meet environmental restoration and waste management needs within the DOE complex. The basic premise was that through development of new technology, cleanup could be achieved "faster, cheaper, better, and more safely" (DOE, 1995b). Since its inception, OTD (now OST) has employed a variety of programs and approaches to fulfill its charge; Table 1 summarizes EM-OST initiatives during the period 1989-98 that are relevant to the DDFA.

To ensure that EM's programs focused on the most urgent environmental restoration and waste management problems, Thomas Grumbly, then Assistant Secretary for Environmental Management, established a Working Group in August 1993 to implement a new approach to environmental research and technology development. A key feature of this new approach was establishment of five focus areas within OTD to address DOE's most pressing problems (DOE, 1995b). These focus areas were:

1. High Level Radioactive Waste Tank Remediation,
2. Mixed Waste Characterization, Treatment, and Disposal,
3. Contaminant Plume Containment and Remediation,[7]
4. Landfill Stabilization,[7]
5. Facility Transitioning, Decommissioning, and Final Disposition[8]

---

[5]The original name was Office of Environmental Restoration and Waste Management; it was renamed in 1994.

[6]EM-50 is one of seven Deputy Assistant Secretary Offices within EM (designated EM-10 through EM-70), all of which report to the Assistant Secretary for Environmental Management (EM-1). OTD (EM-50) was renamed the Office of Science and Technology (OST) in 1995.

[7]Focus areas 3 and 4 were later combined and renamed "Subsurface Contaminants."

[8]First renamed Decontamination and Decommissioning, and more recently Deactivation and Decommissioning Focus Area (DOE, 1998a).

TABLE 1 Major EM-OST Initiatives Since 1989

Program	CY89	CY90	CY91	CY92	CY93	CY94	CY95	CY96	CY97	CY98
Integrated Technology Program	■	■	■	■	■	■				
Integrated Technology Demonstrations	■	■	■	■	■	■				
Focus Areas						■	■	■	■	■
Large-Scale Demonstration Programs							■	■		
EM Science Program									■	■
Ten Year Plan									■	■
Technology Deployment Initiative									■	
Accelerated Site Technology Deployment										■

The focus areas formed the core of OST's integrated team structure for research and technology development to assist EM's mission. As an adjunct to the establishment of the focus areas, the NRC's Committee on Environmental Management Technologies (CEMT) was formed in 1994 at the request of Mr. Grumbly to provide continuing independent advice to DOE-EM on its technology development program. As part of this effort, a subcommittee on D&D was formed under the CEMT.[9]

The DDFA was formed to assure that adequate technologies are available to support EM's task to deactivate more than 7,000 contaminated buildings and fully decommission about 10 percent of them. In addition to the buildings themselves, the task included decontamination of the metal and concrete within those buildings and disposal of some 180,000 metric tons of scrap metal (DOE, 1996b). When the CEMT Subcommittee on D&D was first briefed (November 1994) by DOE personnel, the efforts of the DDFA had been divided into four main activities: facility deactivation, facility decontamination, facility dismantlement, and materials disposition. The subcommittee was given the three draft planning documents that established the DDFA program in 1995.[10] Also in 1995, OST designated the Morgantown (West Virginia) Energy Technology Center (METC) as the lead organization for the DDFA. Now called the Federal Energy Technology Center (FETC), it administers the DDFA programs (Bedick et al., 1996). Table 2 summarizes FETC programs in the DDFA for 1997.

The emphasis in the DDFA was to select and demonstrate technologies that were mature enough to be implemented in actual cleanup activities, i.e., those that were deemed to be ready for the end user. The DDFA intended to look first at commercial technologies that had never been used in the DOE complex or that had been used in the complex but not applied to D&D before developing truly new technologies (DOE, 1996b). The DDFA applies the term "innovative" technology to those technologies that 1) have been used commercially but not yet applied in the DOE environment, 2) are being used in a new way, or 3) are still under development. This approach follows the basic premise within EM that technology development will result in substantial reductions in time and cost and in increased effectiveness of the cleanup work as compared to the baseline case, which assumes the use of "available" technologies[11] (DOE, 1996c). To encour-

---

[9]The D&D subcommittee, as well as the other CEMT subcommittees, were reorganized as independent committees under the NRC Board on Radioactive Waste Management in 1997.

[10]The three documents, in draft form, were: *Decontamination and Decommissioning Focus Area: Technology Summary*, DOE/EM-0253, June, 1995; *Strategic Plan: Decontamination and Decommissioning Focus Area, Pre-Decisional Draft*, July 14, 1995; and *Implementation Plan, Decontamination and Decommissioning Focus Area*, prepared by U.S. DOE Morgantown Energy Technology Center, July 18, 1995.

[11]"Available" technologies are those currently used within the DOE complex. The expected cost and performance "baseline" is derived assuming their use. Such technologies are referred to as "baseline technologies" in this report.

TABLE 2   FETC D&D Program Status at End of
1997 (Hart, 1997)

Program	1997 Budget	Number of Projects
Basic Science	$1.8 million	
Characterization		3
Deactivation		1
Decontamination		5
Containment		2
Waste Reduction		3
Applied R&D	$19.8 million	
Characterization		4
Deactivation		4
Dismantlement		4
Waste Disposition		3
LSDPs	$8.1 million	
Fernald Plant 1		10/4*
Hanford C-Reactor		14/5*
CP-5 Reactor		21/8*

*Technologies demonstrated/technologies deployed.

age DOE sites to use innovative technologies and to develop cost comparison data, the LSDP was established in 1995. The LSDP was intended to be a cornerstone of DDFA and provide a means for side-by-side comparison of "innovative" versus baseline technologies (DOE, 1996b). As specified in the statement of task, the LSDP is a significant part of this committee's review. The program is discussed in detail later in this report.

In 1996, Alvin Alm succeeded Thomas Grumbly as Assistant Secretary for Environmental Management. Mr. Alm announced a *Ten Year Plan*, which stated: "Within a decade, the Environmental Management program will complete cleanup at most sites" (DOE, 1997c).[12]   DOE sites were asked to revise their cleanup strategies and develop plans in response to this new directive. Although it represented a fundamental change in EM's site cleanup schedule, the *Ten Year Plan* acknowledged that even with the accelerated effort, "50 percent of the work" would remain after year 2006, particularly at the larger sites (DOE, 1997c). Budget estimates show that about 40 percent of the total cleanup cost (in 1998 dollars) will be incurred before year 2006 and about 60 percent after year 2006 (DOE, 1998b). Because D&D occurs near the end of the chain of activities in

---

[12]Mr. Alm's initiative originally was referred to as the *Ten Year Plan*, subsequently the *2006 Plan* and, most recently, *Paths to Closure* (DOE, 1998b). Mr. Alm described his *Ten Year Plan* to the NRC Board on Radioactive Waste Management in October 1996. The final version, *Paths to Closure*, was not reviewed by the committee.

each site, and because the most difficult D&D tasks will be encountered at the larger sites, about 80 percent of the total costs of deactivation, decontamination, and decommissioning will be incurred after the year 2006 (DOE, 1997d).

Deploying alternative and more effective technologies is expected to play a key role in the productivity enhancements required by the *Ten Year Plan* (Alm, 1997). Accordingly a new program, the Technology Deployment Initiative (TDI), was established by OST in 1997 to promote deployment of previously developed OST technologies for actual cleanup applications (DOE, 1997c). The TDI is discussed in more detail later in this chapter.

In July 1996 DOE/OST published a D&D technology needs document that was intended to represent complex-wide DOE technology needs (DOE, 1996a). This document was compiled from needs identified by the Site Technology Coordination Group (STCG) that was formed at each DOE operating site. Each STCG consisted of a site coordinator and members who compiled site technology needs in each of OST's focus areas.[13]   Altogether some 535 technology needs were identified, including about 100 needs in the DDFA. Linkage Tables that identified multiple site needs for a given technology subsequently were developed (DOE, 1997a).

In the third quarter of 1996, in DOE's appropriations for FY97, Congress directed that $50 million be used to develop a basic research program focused on long-term cleanup needs across the DOE complex (NRC, 1997). The program, called the Environmental Management Science Program, has the mission to (1) develop targeted, long-term basic research for environmental problems so that breakthrough approaches will lead to significantly reduced cleanup costs and risks to workers and the public; (2) bridge the gap between broad fundamental research and needs-driven applied technology, and; (3) serve as a stimulus for focusing the nation's science infrastructure on critical national environmental management problems (NRC, 1997). Funds provided by the program are awarded competitively to national laboratories, other federal laboratories, and academic and industrial organizations. To ensure that the program is mission oriented and that its achievements are recognized and used by EM, the science program is integrated closely with the focus areas and also is coordinated with the DOE Office of Energy Research to ensure that the broad-based fundamental research and development supported by that office is used (DOE, 1996d). In 1997 basic science program funding for DDFA projects was $1.8 million (Table 2).

THE LARGE-SCALE TECHNOLOGY DEMONSTRATION PROGRAM

The Large-Scale Technology Demonstration Program was intended to be a cornerstone in the deployment of innovative technologies selected by DDFA. In

---

[13]STCG members include representatives of DOE, contractors, the Environmental Protection Agency (EPA), local tribal nations, and other stakeholders.

accordance with its Statement of Task (Appendix A), the D&D committee spent a great deal of time in evaluating the utility and effectiveness of this approach. The stated purpose of the LSDP is to validate performance of D&D technologies and introduce the application of alternative technologies in parallel with baseline technologies (Boyd, 1997b). The demonstration is carried out as a part of an actual, ongoing D&D project at a DOE site. To achieve its purpose the LSDP is intended to:

- Reduce risks from first time application of "new" D&D technologies;
- Provide an opportunity to compare performance of new technology to baseline technology; and
- Provide an opportunity to showcase new technologies and vendors.

The LSDP was initiated in July 1995 when FETC sent a Request for Letter Proposals to all DOE Operations Offices. The request contained the LSDP selection criteria established by the DDFA. These criteria included: significance of the proposed demonstration (especially for cost reduction in future projects), readiness of the proposed demonstration, site commitment, and project management (DOE, 1996b). The sites were requested to offer facilities that were undergoing D&D as a part of the site operation to be host facilities for demonstrations of new technologies. This would afford opportunities to test the innovative (new to the DOE complex) technologies alongside baseline technologies. Eight letter proposals were received in response to the request. In October 1995 DDFA selected proposals from three sites to incorporate LSDPs into actual D&D projects. These projects were the Plant 1 uranium processing complex at the Fernald Environmental Management Project in Ohio, the CP-5 Reactor at Argonne National Laboratory in Illinois, and the C-Reactor at the Hanford Reservation in Washington (DOE, 1996b). During this study, committee members visited these locations to observe the LSDP activities in progress and to discuss the demonstration with DOE and contractor personnel on site. Reports of these site visits are given in Appendixes C-F of this report.

The initial LSDPs had several common elements: Decommissioning plans using standard (baseline) technologies already had been established, technology selection criteria had been established (albeit independently at each location), industry-government partnerships were in place, and, most importantly, each site had requested that it participate. Technologies actually demonstrated have come mainly from the commercial sector. Table 3 provides a list of technologies that were demonstrated during the committee's review period.

Cost savings versus the baseline technology were expected to provide incentive for the DOE sites to adopt innovative technologies. In June 1995 the DDFA acquired the services of the U.S. Army Corps of Engineers (the Corps) to establish an independent comparison of the cost of accomplishing a task using an innovative technology versus the baseline technology (Kessinger and Greenwald, 1997). Initially DDFA planned for the Corps to use baseline costs supplied by the site, and independently estimate the cost of using the new technology to accomplish the

TABLE 3 LSDP Technology Demonstrations

A. Technology Demonstrations at CP-5 Reactor

Technology	Problem Area	Description
Mobile Automated Characterization System	Characterization	Battery-powered, autonomous robot base with a laser positioning system that can detect alpha and beta/gamma contamination.
Pipe Explorer	Characterization	In-situ piping characterization system.
X-ray Fluorescence Detection	Characterization	In-situ characterization system.
Gamma Camera	Characterization	Visual imaging of area radiation levels.
SRA Surface Contamination Monitor	Characterization	Consists of a wide area, computer controlled, position sensitive proportional counter that is mounted on a motorized cart (used at CP-5 for beta/gamma).
Pipe Crawler	Characterization	In-situ piping and duct work characterization system.
Field Transportable Beta Counter	Characterization	Instrument provides for real-time detection and spectral analysis of Sr-90, Cs-137, Tc-99, and other beta emitters in the 40-picocurie range.
In-situ Object Counting System	Characterization	Radiological monitoring system used to measure small levels of contamination on large objects or surfaces.
Pegasus Coating Removal System	Decontamination	Coating removal system using strippable chemicals.
Centrifugal Shot Blasting	Decontamination	Effectively removes layers of concrete to varying depths, without dust.
Flashlamp Decontamination	Decontamination	Intense light breaks the chemical bond between the material and surface. Material residues and gases are collected by vacuum.

Rotopen Scabbling	Decontamination	This decontamination system for concrete surfaces consists of abrading media (Heavy Duty Roto Peen), surface planing equipment, and high-volume vacuum.
Concrete Milling	Decontamination	Permits selective removal of contaminants from concrete substrate.
Advanced Recyclable Media System	Decontamination	Blasting system using recyclable, specially impregnated sponges.
Empore Membrane Separation	Decontamination	Uses ion-exchange resins embedded in membrane material within a net-like fibril matrix to remove contaminants in water.
Swing-Free Crane	Dismantlement	Allows a load suspended from a gantry crane to be moved without inducing any undesired swinging motion.
Dual-Arm Work Platform	Dismantlement	Provides flexibility for cooperative and coordinated actions by using two robotic arms.
Rosie Mobile Work Station	Dismantlement	An electro-hydraulic, omni-directional locomotor platform with a heavy manipulator mounted on its deck. The heavy manipulator boom can deploy a large number of tools for demolition and decontamination.
Remote Controlled Concrete Demolition System	Dismantlement	Uses a remote-controlled, track-driven, service-robot, known as the Brokk BM 150, that employs an articulated hydraulic boom with various end-effectors to accomplish a variety of tasks.
NU-FAB Suit	Health and Safety	One-piece, microporous, disposable, waterproof coverall to be used in hot/wet atmospheres.
FHRAM-TEX Cool Suit	Health and Safety	One-piece disposable, breathable, waterproof coverall for hot/wet atmospheres.

Continues

TABLE 3   Continued

B. Technology Demonstrations at Fernald Plant-1

Technology	Problem Area	Description
Field Raman Spectroscopy	Characterization	A fast and effective way to detect contaminants through in-situ compound analysis.
Laser-Induced Fluorescence Imaging	Characterization	A fast and accurate uranium characterization tool.
Pipe Inspection System	Characterization	Involves a small monitor/VCR equipped with a tiny, light-bearing camera probe used to perform remote inspection and record results.
Sponge Jet Cleaning of Equipment	Decontamination	This technology is comparable to steam-jet cleaning or water washing technologies. However, since it does not use water, it can be used to clean material containing enriched uranium.
Steam Cleaning with Vacuum Recovery	Decontamination	This technology uses a pressurized, heated stream of water that flashes to steam when it impacts the surface being cleaned. The steam is then vacuum collected and recycled.
Oxy-gasoline Torch	Dismantlement	Low-cost, clean-cutting, highly effective metal cutting technology.
Vacuum Removal of Insulation	Dismantlement	System was used to remove rock-wool insulation efficiently from walls while controlling airborne contamination.
Void Filling with Low-Density Cellular Concrete	Waste Management	Used to fill voids in hollow components to meet waste disposal cell acceptance criteria.
Void Filling with Foam	Waste Management	Used to fill voids in hollow components to meet waste disposal cell acceptance criteria.

Technology	Problem Area	Description
Personal Ice Cooling System	Health and Safety	Incorporates small diameter tubing into a comfortable full-body suit. Chilled water is circulated through tubing that is attached to a pump unit and a small ice container.

C. Technology Demonstrations at C-Reactor

Technology	Problem Area	Description
Laser-Assisted Ranging and Data System	Characterization	Performs accurate and repeatable radiological characterization surveys of indoor building surfaces. It is electronically coupled to a data collector (AutoCad).
Gamma Ray Imaging	Characterization	Provides a map of a dose rate of an area superimposed on its visible image.
Position Sensitive Radiation Detector/Monitor	Characterization	This detector design turns one large gas-flow proportional counter into 400 or more accurate and sensitive mini-detectors. (Used for beta/gamma and alpha.)
Concrete Shaving	Decontamination	This concrete decontamination system has shaving blades that remove precise layers, leaving a completely smooth, finished surface.
Surface Decontamination*	Decontamination	Will test the ability of innovative technologies to remove surface contaminants from lead, concrete, and the asphalt emulsion covering the fuel basin.
Structural Steel Decon/Recycling*	Decontamination	Will test the ability of innovative technologies to process and free release structural steel for reuse/recycle.
Self-Contained Pipe Cutting Shears	Dismantlement	Battery powered, hand-held, hydraulic mini cutter used for cutting 1"-2" small bore piping.
Large Bore Pipe Cutting*	Dismantlement	Equipment/process that can cut large-bore, horizontally mounted pipe in a congested area, with or without asbestos, which is lightweight and generates low heat.

Continues

TABLE 3 Continued

C. Technology Demonstrations at C-Reactor

Technology	Problem Area	Description
STREAM Management Database System	Health and Safety	STREAM (System for Tracking Remediation, Engineering, Activities & Materials) is a Management Database tracking system that provides a number of advantages such as visual and audio training, characterization information, waste tracking and manifests, and tracking of workers training and exposure records.
Mobile Integrated Temporary Utility System	Health and Safety	Integrated temporary power, communication, safety, and alarm systems.
Self-Contained Air Cooled Respirator/Suits	Health and Safety	Liquid air, self-contained breathing and cooling system with duration of 2 hours; worn as a backpack.
Heat Stress Monitoring System	Health and Safety	On-line human-monitoring system developed to provide monitoring where heat stress or other physiological safety issues are a concern.
Sealed Seamed Sack Suits	Health and Safety	Lightweight, durable, waterproof and breathable protective coveralls are assessed against baseline cotton coveralls.
Reactor Stabilization*	Stabilization	Spray-on coating, surface encapsulation to stabilize contaminants on reactor face.

*Demonstrations employ the use of a broad industry search to identify technologies that address these specific problems.
Table courtesy of Steve Bossart and Ken Kasper of DOE FETC, based on their tables from the article, *Improved D&D Through Innovative Technology Deployment*, published in *RadWaste Magazine*, January 1998. Reproduced with authors' permission.

same task, based on its actual use in the demonstration. Subsequently, the Corps found that to put the costs on a comparable basis it was necessary that it also estimate the baseline costs (Kessinger, 1997). Further, the Corps found it necessary to average cost standards for labor rates and disposal among the larger sites to obtain a reasonable basis for the cost of implementing a new technology at sites other than the host site (U.S. Army Corps of Engineers, 1997).

As the program got under way in August 1996, the DDFA stated that these first three LSDPs would be completed by the end of calendar year (CY) 97, and an additional five demonstrations would be completed by the end of CY99. By the end of fiscal year (FY) 2000, DDFA planned, by way of these eight LSDPs, to have demonstrated technologies, systems, and methods to provide capabilities to D&D 90 percent of the surplus facilities and materials (Bedick et al., 1996). Information on the results of the demonstrations was to be promptly disseminated to encourage deployment of successful new technologies. This was to be accomplished primarily by short, one- or two-page fact sheets describing the technology and by detailed summary reports (referred to as "Green Books") that would include details of the demonstration and the Army Corps of Engineers' cost comparison data.

Near the end of the committee's review period, in November 1997, the 90 percent demonstration goal had slipped two years, to 2002. Two additional LSDPs scheduled to begin in 1997 had been canceled by the DOE field offices. These were the Rocky Flats Building 779 Complex, which was canceled in August 1997, and Oak Ridge Building K-27, which was canceled in October 1997[14] (DOE, 1997f). By the end of its review period in December 1997, the committee had not received any of the expected technology summaries, and it was understood that none of the three initial LSDPs had been completed.[15]

TECHNOLOGY DEPLOYMENT INITIATIVE

OST has been criticized by both Congress and the General Accounting Office (GAO) for a poor track record in deployment of new technologies.[16] Resistance from the operating sites themselves was seen by OST as a major barrier to deployment of new technologies because the operators considered the new technology to be "risky" or simply "not invented here" (GAO, 1996, 1997). Seeking

---

[14]Four new Large-Scale Demonstration and Deployment Projects (LSDDPs) were announced in 1998 as this report was being completed: Transuranic Waste Technologies at Los Alamos National Laboratory, Dismantlement of Tritium-Contaminated Facilities at Mound, Deactivation of 321-M Fuel Fabrication Facility at the Savannah River Site, and Fuel Storage Canals at the Idaho National Environmental and Engineering Laboratory .

[15]Five of about 50 expected Technology Summaries were published in February 1998. Since that time, additional summaries have been published.

[16]Hearings conducted by Congressman Tom Bliley (R-Va.), Chairman of the House Commerce Committee, and GAO audits and testimony (GAO, 1996, 1997).

to overcome this resistance, OST began the Technology Deployment Initiative in early 1997.

The TDI mission is to deploy technologies and processes that reduce the DOE-EM "mortgage," accelerate site cleanup, and enhance achievement of *Ten Year Plan* goals (DOE, 1997g; Alm, 1997). Deployment is defined by OST as implementation of a technology or process through *multiple* sites or applications. To accomplish this mission, TDI provides financial incentives to sites to implement fully developed, previously demonstrated innovative technologies or processes into their existing or planned cleanup activities. OST has acknowledged that accomplishing the TDI mission will require DOE complex-wide cooperation (DOE, 1997g).

The TDI has the following objectives:

• Provide for accelerated multiple deployment of technologies or processes to conduct cleanup (including decontamination and decommissioning) in ways that will reduce EM cleanup costs and support *Ten Year Plan* goals;

• Provide third-party validation of cost savings and incentives to site participation through reinvestment of cost savings;

• Break down barriers that inhibit implementation of technologies or processes; and

• Achieve closer coordination through joint ownership and funding of projects across DOE EM organizations (DOE, 1997g).

The TDI was to be a multi-year program with an initial funding level of $50 million for FY98. Proposals from DOE field offices were sought for multiple deployment of technologies and processes for site cleanup (including environmental restoration and waste management, as well as D&D). A screening criterion for all proposals was that they show significant savings in life-cycle costs compared with the baseline technology and that deployment throughout the complex be accomplished without additional OST funding. Proposals were envisaged to not exceed $5 million each unless an exceptional return on investment could be shown (DOE, 1997g; Hyde, 1997). By funding the initial deployment of new technologies, OST intended to alleviate site concerns about incurring technological risk or penalties for schedule delays.

As a part of OST's increased emphasis on technology deployment assistance, the TDI program was renamed the Accelerated Site Technology Deployment (ASTD) Program in early 1998, as this report was being finalized. The DDFA announced three ASTDs for 1998: 1) Enhanced In Situ Decontamination and Size Reduction of Glove Boxes at Rocky Flats, 2) the Los Alamos National Laboratory (LANL) Decontamination and Volume Reduction System, and 3) the Idaho National Environmental and Engineering Laboratories (INEEL)/Fernald Environmental Management Project (FEMP) Integrated Decontamination and Decommissioning. Because the TDI/ASTD began near the end of the committee's review period, it is not assessed in detail in this report.

2

Committee Findings

In its statement of task, the D&D committee was asked to review the approaches used by the DDFA to assess their utility and effectiveness. The review was to include the DDFA's means of selecting and evaluating alternative technologies and its methods of seeking deployment of new technologies, especially the LSDP. In addition the committee was to review the processes that DDFA is using to compare the advantages expected from technology development to existing (within DOE) processes.

To accomplish its task, the committee held a series of fact-finding, discussion, and writing meetings, as described in Appendix B. Some of the committee members also visited the sites hosting the LSDPs. Reports of these site visits are given in Appendixes C-F. The committee's findings are based on literature made available by OST and the DDFA during the review, presentations at its meetings, and the site visits. Findings relevant to the statement of task are discussed in six topical areas:

- Strategic planning,
- Evaluation and prioritization,
- End-state specification,
- Dissemination of results,
- Technology assessment, and
- Technology implementation.

GENERAL FINDINGS

Since its inception, OST has been exposed to a great deal of external criticism focused primarily at lack of significant new technology deployment com-

19

mensurate with the amount of money spent (GAO, 1996, 1997). The OST responded to this criticism by an almost continuous change in programs, processes, and stated priorities. These are discussed in Chapter 1 and summarized in Table 1 (p. 7).

Within the DDFA, the committee found that continual superficial changes, for example, changing the names of its key programs (the Technology Deployment Initiative is now the Accelerated Site Technology Deployment Program, and the Large-Scale Demonstration Program became the Large-Scale Demonstration and Deployment Program) was detrimental to the DDFA's effectiveness. Such changes gave the impression that the programs were *ad hoc* and lacked follow-up. This was symptomatic of the lack of strategic planning, prioritization, specification of objectives and documentation described in the following sections of this chapter. By implying significant changes in program content when there were none, DDFA undermined its credibility with its potential DOE customers and industrial partners. In addition to the perception of DDFA's programs as a "moving target," the committee found the complexity of DOE purchasing, contracting, and other procedures to be a significant inhibition to participation in DDFA programs by outsiders. Greater participation by private industry and universities might have provided a greater diversity of innovative technologies to the LSDPs. The DDFA's Requests for Proposals for the LSDP were addressed only to DOE offices and not to the commercial sector. Information on potential industrial partnerships for the LSDPs appears to have been available mainly to contractors already familiar with the sites. For example, there was no notice published in *Commerce Business Daily* when the Strategic Alliance for CP-5 was being formed (Appendix C).

In response to external criticism and budgetary pressure, however, recent efforts by OST have been promising. The committee acknowledges that these recent efforts (described by Gerald Boyd in OST's final presentation [Boyd, 1997b] to the committee) to formalize the prioritization process, to involve technology users, and to improve communications and information dissemination, represent necessary and long overdue improvements in OST and the DDFA. These efforts are still in the planning stages, and it is too early to judge whether they will be successful. The committee can only comment on programs that were in place during the review period covered by this report, which ended in December 1997.

STRATEGIC PLANNING

Implementation of the DDFA program has suffered from the lack of a current strategic plan on which to base discussions, select alternatives, and manage the program. The planning documents provided to the original CEMT D&D committee in 1995 were never issued in final form, nor were revised planning documents provided during the committee's review.[17] Numerous changes in policy, objectives, schedules, and funding since 1995, especially the *Ten Year Plan* with its increased emphasis on technology deployment, have made previous DDFA planning obsolete (DOE, 1997c). Up-to-date planning documents were deemed necessary by the committee to define an ordered and structured sequence of activities for DDFA's technology development and implementation. In the absence of an up-to-date OST-DDFA strategy, new programs were initiated without documented evaluation of their expected results or of their impacts on existing efforts. The increased number of DDFA-related initiatives by EM-OST during 1996-97 is shown in Table 1 (p. 7). Shifts in approach, such as the emerging emphasis on deployment on top of the existing efforts at technology demonstration (the LSDPs), caused competition among new and existing programs for limited resources. As a result, DDFA has experienced increased difficulty in meeting its previously announced objectives (for example, the continued schedule slippage of the LSDPs and their documentation discussed in Chapter 1).

A conclusion of the 1995 CEMT report was that DOE should address planning in terms of a process, since "many organizations have found that the most valuable aspect of any strategic planning exercise is the process of assembling the plan rather than its specific details" (NRC, 1996). The CEMT report recommended that the planning "receive the undivided attention of the highest-level decision makers." In the committee's view, a successful plan should provide sufficient guidance for technology development to ensure the timely availability of technologies needed to achieve a specified end state for each facility undergoing D&D. It would include analysis of the capabilities of existing technologies and provide for development of new technologies where existing technology is lacking or where a new approach could perform faster, cheaper, and more effectively. Such strategic planning for the DDFA has not been accomplished.

Near-Term (Pre-2006) Strategy

The current DOE approach to the cleanup of its radiologically contaminated weapons complex facilities is to make maximum use of existing and proven techniques to achieve as much cleanup as practicable in the shortest time. The stated goal of EM's *Ten Year Plan* is to clean up 90 percent of its sites within the next 10 years (DOE, 1997c, 1998b). This is a commendable goal, but it clearly

---

[17]The documents, in draft form, were: *Strategic Plan: Decontamination and Decommissioning Focus Area*, Pre-Decisional Draft, July 14, 1995; and *Implementation Plan, Decontamination and Decommissioning Focus Area*, prepared by U.S. DOE Morgantown Energy Technology Center, July 18, 1995.

leaves the more difficult, the more intractable, and the more costly cleanup matters for later (long-term) treatment (DOE, 1997c, 1997d).[18]

For near-term D&D projects, the committee agrees with DDFA's approach of identifying and validating, by on-site demonstrations, existing technologies and commercially available practices that are currently not used on DOE sites. However, information presented to the committee indicated that capabilities already residing in the private, academic, and foreign sectors were not being incorporated efficiently into the DDFA's short-term planning. In fact it became clear during several question sessions that such sources were mostly untapped. In cases where these "new" technologies demonstrably are superior to baseline or "available" technologies, they can be applied successfully to enhance EM's near-term D&D capability. Further, the committee believes that there is insufficient time to develop, test, and apply unproved, truly innovative, technologies to effect major cleanup tasks by 2006.

OST's shift in emphasis from technology development toward technology deployment assistance is consistent with accelerated cleanup. If the expected benefits from innovative technologies proposed by the DDFA are to be realized, it will be necessary for DDFA to achieve greater effectiveness in technology deployment than was seen by the committee during its review.

Long-Term (Post-2006) Strategy

While the emphasis of the *Ten Year Plan* is on proceeding as quickly as possible with actual cleanup work, the plan also considers development of advanced technologies necessary, both to address cleanup tasks not amenable to available techniques and to reduce costs and risks of the cleanup program. The committee believes that ten years or more is a realistic time frame for development, demonstration, and deployment of truly innovative (i.e., unproved or presently unknown) technologies.

The committee found that an opportunity exists now for the DDFA to develop an effective long-term strategy that leads to the deployment of advanced technologies after 2006. It seems prudent for the DDFA to concentrate its technology development activities on addressing longer-term needs. To focus this activity, both site-specific and complex-wide problems that are either intractable or very difficult (e.g., expensive) with current technologies must be identified and prioritized. An effective strategy will target those problems with the highest complex-wide return on investment and include the research and development

---

[18]It is noted that about 60 percent of the total cleanup *cost* will be incurred after 2006. For D&D activities, coming generally near the end of the overall program, it has been estimated that 80 percent of the costs will be incurred after 2006.

capabilities of industry, the national laboratories, and research universities as appropriate. A review of OST's programs, including the DDFA, by the Strategic Laboratory Council (composed of one representative from each national laboratory) presented suggestions for greater use of the national laboratories as sources of new technologies, especially through the EM Science Program (SLC, 1997). Because D&D is a growing need for all industrialized countries, non-U.S. programs are another source of technical innovations.

EVALUATION AND PRIORITIZATION

The *sine qua non* of effective technology development is the establishment of specific needs that the technology is to meet. The "National Needs Assessment" compiled from needs identified by the STCGs and issued in July 1996 was a significant step in this direction (DOE, 1996a). This document, which was intended to represent the complex-wide DOE technology needs, resulted in the identification of 102 D&D needs. These were grouped into 31 categories. In several of the DDFA's presentations to this committee, site needs were listed as a criterion for selecting technologies to be demonstrated in the LSDPs; however, no presentation or publication from the DDFA has shown a clear relationship between these categories and the technologies selected for the LSDPs or TDIs.

The committee finds that effective overall evaluation of the site needs and prioritization of those needs is missing from the present DDFA approach. In addition to the initial "bottom-up" input from the STCGs, "top-down" prioritization for program planning is necessary. Such an exercise requires the judgment and participation of senior D&D subject-matter experts both external and internal to DOE, who can provide input on existing commercial and other government agency technologies and thus develop an overall, prioritized, top-down evaluation of needs to feed into the DDFA program. The committee also believes that the emphasis of this effort should be to identify and prioritize only those technologies needed beyond 2006 that may provide significant cost and/or safety benefits in the long term, as well as those few problem areas where no technology currently exists.

END-STATE SPECIFICATION

The 1995 CEMT report stated that "DOE is in urgent need of defining criteria by which the level of site cleanup on a 'necessary and sufficient' basis within regulatory constraints can be determined" (NRC, 1996). This level of cleanup, or "end state" is the final condition of the building, facility, or site after cleanup is accomplished. In the case of D&D, an agreed upon end state may require little cleanup for a building planned for re-use by DOE, or it may require demolition of the building and return of the land to unlimited use. End states must be known and specified in order to develop a realistic and effective cleanup program. Determination of end states for DOE sites or facilities is clearly beyond the authority of OST and is not even entirely within the province of DOE, considering stakeholder and regulatory agency involvement. Nevertheless, the committee would have expected the DDFA to use its best judgment to define reasonable target end states as a standard for determining needs for new technologies, comparing technologies, and measuring their degrees of success. In its present review, the committee found no evidence of significant progress in the DDFA in defining or even proposing end states to be targeted by its new technologies, nor did there appear to be any sense of urgency in the DDFA to address this need. The committee believes that formulating a coherent program to develop new D&D technologies cannot be done unless the target end states to be achieved are specified.

Without defining "how clean is clean enough" for a given D&D task, the technology needs, costs, and schedules cannot validly be established. The defined end state also determines characteristics of secondary waste from the D&D activity. Volume, radioactivity, and other chemical and physical properties of the waste stream will determine recycling and disposal options, cost, and, ultimately, whether the proposed D&D technology is adequate. The technology required to reach one level of residual radioactivity may be entirely different from the technology required to reach a different level, either higher or lower. Depending on the required end state, there may or may not be a need for developing a new technology to meet the requirement.

By developing reasonable target end states, DDFA will be able to justify its technology development program better. For example, presently available technology may not be able to reach a target end state, in which case, improved technology clearly is needed. Alternatively, in seeking to deploy a new "cheaper and faster" technology, demonstration that it can achieve the same end state as the baseline technology is essential. As in a previous report (NRC, 1996), the committee notes that before undertaking any technology development program, it is essential to define and specify the particular problem that is to be solved, rather than to develop a solution and then look for a problem for which the technology can be used.

DISSEMINATION OF RESULTS

OST does not have direct responsibility or authority for implementing or using the technology it demonstrates. Before the DOE field offices or site contractors will be willing to use its technology, OST/DDFA must sell the technologies through effective communication. Potential users must be aware not only of the technical results of the DDFA's developments and demonstrations but also of the factors that make a given technology more attractive, or less attractive, than the baseline technology. The lack of success in gaining user support for its initiatives that the committee observed during its review period indicates that the DDFA is not communicating effectively with its customers.

As described in Chapter 1, the Innovative Technology Summaries or "Green Books" were planned to be the primary vehicle for disseminating results of the LSDPs. DDFA planned to describe each successful technology demonstration in sufficient detail, including comparative cost data, to convince potential users complex-wide to adopt the new technology in place of a previous baseline technology. It is a significant deficiency that no Green Book had been published by the end of the committee's review period, December 1997. Several technologies had been demonstrated more than a year previously.[19]   Only five of an expected total of about 50 Green Books had been published as of February 1998. Although in his final presentation to the committee, Gerald Boyd stated that of 39 demonstrated technologies, EM-40 had adopted 14 without requiring Green Books (Boyd, 1997a), the committee found that DDFA has fallen short of its initial intent to disseminate information resulting from the LSDPs promptly.

In addition to timely dissemination of results, the Green Books must provide important new information to DDFA's customers. However many of the technologies demonstrated in the LSDPs that will appear in Green Books are simply off-the-shelf commercial products. Some examples are the cutting and surface ablation techniques listed in Table 3. The committee found that DDFA may be harming its own credibility by promoting these as innovative technologies, and it questions whether simply testing and providing information about commercial technologies is a suitable role for the DDFA.

TECHNOLOGY ASSESSMENT

With increasing demand for justification of OST programs, methods must clearly be defined for determining the potential advantages of an innovative technology over the baseline.[20]   Although the potential for direct cost savings in the D&D operation is of primary importance, there are other technical and institutional issues that can have a major impact on the decision process.

A key feature of the LSDP was to be the direct, side-by-side comparison of baseline and new technology, as discussed in Chapter 1. During the site visits,

---

[19]For example, the Kelly Steam/Vacuum decontamination system, the AEA High-Pressure Water Sponge Jet, and the Vector Technologies VecLoader for asbestos removal were demonstrated at Fernald in August 1996. Green Books for these technologies had not been published as of April 1998.

[20]Hearings by the Bliley committee and GAO audits were noted in Chapter 1. Testifying before the Bliley committee on May 7, 1997, Dr. Edgar Berkey, Chairman of the Committee on Technology Development and Transfer of DOE's Environmental Management Advisory Board (EMAB), stated that one of his committee's findings was that better performance metrics for EM's technology program are likely to yield better results (Berkey, 1997).

committee members discussed the procedures for selecting baseline and innovative technologies for each LSDP with DDFA and site personnel (Appendixes C-F). These discussions indicated that the technology selection procedures were not consistent from site to site. In addition, the DDFA's baseline technology, against which its innovative technology was to be compared, was not necessarily the technology used by the site contractor for the actual D&D of the facility. In some cases the baseline technology was not demonstrated as a part of the LSDP at all, but was a technology that had been used elsewhere. The committee found that actual side-by-side comparisons of technologies were achieved in a convincing manner in relatively few cases. The LSDPs therefore failed to provide the solid basis for technology comparison and assessment promised at the program's inception.

Cost Estimation

Since only technologies that have been proven to be safe and meet regulatory criteria can be considered for implementation, the choice among technologies is based almost entirely on economics. During its review, the committee placed considerable emphasis on evaluating the DDFA's methods for estimating cost savings that could result from implementing innovative technologies versus the baseline technologies. Common costing and performance assessment methodologies are essential if comparison of baseline and innovative technologies is to be done. The committee acknowledges that the DDFA has made a conscientious effort to provide accurate cost estimates of technologies tested in the LSDPs through its agreement with the U.S. Army Corps of Engineers (see Chapter 1).

When the Corps first described its role to the committee in June 1997, the committee felt that the task assigned to the Corps was insufficient to provide reliable comparative cost estimates. Specifically, the Corps was not tasked with developing cost estimates of the baseline technology, but rather was to accept these data from DOE. Subsequently in the course of developing its comparative cost estimates, the Corps found it necessary to perform cost estimates of both the baseline and the innovative technology (Kessinger, 1997). The Corps therefore requested a change in its scope of work and will provide all cost estimates that are to be used in the DDFA's Green Books.

The committee, however, has reservations about the reliability of the comparative cost estimates and believes that future technology demonstrations in the LSDPs (as well as new deployments in the TDIs) can be planned and conducted to provide better data for cost estimation. The committee's reservations arise from three observations: first, many of the baseline technologies were not actually demonstrated;[21] second, the scope of the baseline and the innovative tech-

---

[21]This is acknowledged in four of the five Green Books that were published in February 1998, which state that the baseline costs are not based on observed data.

nology tests was different in some cases; and third, cost and bidding procedures differ among the DOE sites (U.S. Army Corps of Engineers, 1997). In the 1995 CEMT report, the committee recommended that DOE base its estimates of cleanup costs on a standard methodology and explicit criteria that would prevail across all DOE programs, sites, and projects (NRC, 1996). Such a common basis is essential for valid economic comparison of alternative technologies.

Other Considerations

In addition to comparative cost, there are other factors that affect the selection of a new technology. These factors can have an overriding impact on the choice of the appropriate technology to accomplish a specific D&D task. The more important factors include technological risk, regulatory acceptance, conflicts with previously established interagency and state agreements on environmental impact, likely end-user acceptance, waste management implications, schedule impact, and D&D worker/public safety. New technologies may be at a disadvantage compared to the baseline just because their use could require altering contractual requirements. Secondary waste produced by a new technology may lead to difficulties since it may well be different from that of the baseline technology and could require a new risk assessment and permitting process. The committee found that the LSDP fails to address these important institutional barriers to the deployment of new technologies.

TECHNOLOGY IMPLEMENTATION

The introduction of new methods and processes understandably brings some apprehension compared to the utilization of older, familiar approaches especially in the larger DOE sites that have well-established procedures and customs for their operation. There are also the realities of schedules, regulations, and agreements with stakeholders that constrain site operators. The committee, as well as others, recognize the reality of the "not invented here" syndrome. The LSDP has fallen behind its original, ambitious schedule and in general has not met with the success initially expected. This lack of success is a clear signal that voluntary acceptance of new technologies by those responsible for implementing the actual cleanup will not occur until the user site organizations are convinced that there are definite advantages in doing so. Deployment strategies so far generally have failed to demonstrate the advantages of using the new technologies.

The case in point is the LSDP that was intended to introduce and demonstrate innovative D&D technologies at full scale alongside existing baseline commercial technologies. Unfortunately, the concept of demonstrating new technologies came to individual cleanup programs late in their schedules and work programs. D&D work at Fernald and at CP-5 had begun well before the idea of testing new technologies at these locations was put forth. As a consequence, the D&D plans previously established for these existing projects had to be changed, schedules revised, and budgets reallocated. In many cases the baseline technology specified for the LSDP differed from that in use by the contractor and was never itself demonstrated. In these cases, valid comparison of the baseline and innovative technologies was not possible. In addition, cancellation by field offices of the two LSDPs scheduled to begin in 1997 was seen as a clear and negative signal by the committee.

To encourage site managers to accept new technologies, OST established its TDI in 1997 to provide financial incentive to potential users of its selected technologies. By the end of its review period, the committee had not been made aware of any documented technology deployment, nor could the committee find evidence to suggest that limited financial assistance will achieve the acceptance that eluded the LSDP. Although the committee felt that it was premature to formally evaluate the TDI (now called the Accelerated Site Technology Deployment Program), the committee expresses reservations about this approach and notes that similar concerns have been raised by other review groups (DOE, 1997d; GAO, 1997).

3

Recommendations

The D&D committee recognizes the difficulties that the DDFA encountered during its first four years (1994-1997) in organizing itself and its programs to meet DOE's needs for D&D technology development. The committee also recognizes the recent efforts described by Gerald Boyd, Acting Deputy Assistant Secretary for Science and Technology, and others in improving the efficiency and deployment record of the DDFA (Boyd, 1997). As a part of that effort this committee was convened to assess the utility and effectiveness of the approaches and processes used by DDFA, especially the LSDP.[22]

During the committee's review, the LSDP lacked planning and did not meet its schedules or goals. Prompt dissemination of sufficient technical and cost data to encourage site managers to adopt successfully demonstrated technologies was not achieved, nor were the LSDPs able to overcome institutional barriers to new technologies. Widespread user acceptance of the demonstrated technologies did not occur during the committee's review period. Therefore, the general recommendation of the committee is that OST and DDFA substantially revise the Large-Scale Demonstration Program or phase it out.

To assist the DDFA in improving its efficiency, especially as it revises the LSDP and makes further efforts to achieve greater deployment of new D&D technologies, the committee makes the following recommendations:

1.   The DDFA should improve its strategic planning. As discussed in the CEMT committee's report (NRC, 1996), a comprehensive strategic plan, with

---

[22]The statement of task for the committee is given in Appendix A.

29

specific objectives and goals, is essential for decision-making in successful management of the DDFA. A high priority should be assigned to updating the 1995 draft Strategic Plan to reflect DOE's current priorities, scope, schedule, and budget. The plan should be disseminated widely to senior managers to provide a common basis for development and use of associated management and implementation plans. The Strategic Plan should be updated and reissued periodically as DOE policies, procedures, and objectives evolve.

2.   **Top Management in OST should evaluate and prioritize the technology needs of the operating sites.** The D&D technology needs identified in each field cleanup planning process must be prioritized and communicated from each site up to OST. After verification and evaluation of actual, as opposed to perceived, technology gaps that cannot be satisfied by existing technology, OST must prioritize the remaining candidate projects for implementation within the constraints of the available budget. This is a *"top-down"* management function and cannot be delegated. This is absolutely essential to ensure that technology project selection will yield an advantageous return in cost, schedule, and personnel safety.

3.   **OST and the DDFA should link all actions and funding to the prioritized needs.** All actions (selection of technologies to be demonstrated, implementation of demonstrations, establishment of rankings for budgetary purposes) and funding by the OST must be supported by "top-down" prioritized actual needs of D&D cleanup projects in progress or scheduled for implementation.

4.   **The DDFA should define a reasonable target end state for each D&D activity.** The DDFA must establish clear performance goals for any new D&D technology to achieve an effective technology development program. In order to establish performance goals, the DDFA should take the initiative to define and propose end states that would be reasonable for specific DOE D&D activities.[23] These steps are necessary to provide a justification for the DDFA to develop new technologies (where baseline technologies cannot reach a specified end state) or to benchmark new technologies that are claimed to be "faster, cheaper, and better" than the baseline. All proposed demonstration projects should be reviewed by the DDFA to ensure that definition of the desired end state for each demonstration project is clear, complete, and consistent with latest changes in DOE strategic plans and negotiated site planning and operations.

5.   **The DDFA should improve its approach to introducing and gaining acceptance of demonstrated technologies.** The LSDP was designed to intro-

---

[23]The concept of end-states is described in Chapter 2.

duce and gain acceptance by site managers of innovative technologies into D&D activities within the DOE complex. Each site already has established methods for performing D&D activities, and sites appear reluctant to take on the perceived risk of adopting alternative methods. OST "marketing" representatives responsible for promoting these technologies to the candidate sites must understand fully the potential needs of users and work closely with the users to successfully resolve their concerns before technology deployment can be a reality. New technologies must be "pulled" by the site cleanup project manager with a problem; they cannot successfully be "pushed" by the technology supplier. Improved approaches to win widespread DOE site acceptance of new technologies, including those given in the next four recommendations, should be implemented by the DDFA. Similarly OST and the DDFA should take a fresh look at the TDI program to determine the effectiveness of financial incentives as a means of obtaining technology deployment.

6.   **OST and the DDFA should develop and apply a uniform and consistent approach to comparative technology assessment across all projects.** To gain credibility with the site managers and other stakeholders, OST should develop and base its comparative technology assessments on a standard methodology that prevails across the various programs, sites, and projects. The committee recommends that DDFA refine its cost estimating methodologies for baseline and alternate technologies so that cost comparisons are meaningful and can be documented fully. Methodologies for incorporating non-economic criteria (safety, human factors, waste generation, degree of maturity, and technological risk) also should be standardized.

7.   **The DDFA should be more aware of technologies developed in the private, academic, and foreign sectors.** The DDFA should develop a well defined and effective procedure to identify and disseminate information on technologies commercially available in the United States and abroad that can be brought to bear on D&D problems within the DOE complex. To achieve this the DDFA should increase its interactions not only with the national laboratories but also with private industry and international organizations, develop more regional diversity in its contacts with universities, and make its technology needs and programs more visible and comprehensible to private industry.

8.   **The DDFA should communicate its program results in a more effective and timely manner.** Failure to provide adequate communication of the results of the demonstrations, tests, or assessments to prospective end users in a timely manner and in sufficient detail greatly reduces the prospects for acceptance and deployment of new technologies. DDFA should change or improve its present approach (the "Green Books") in order to assure the timely issuance of complete and defensible detailed results.

9. The DDFA should establish a better connection between university and industry programs and prioritized long-term needs. As part of its long-term strategy, the DDFA should become more familiar with programs sponsored by or in progress at universities, industrial organizations, and other government organizations that may be applicable to D&D activities. The DDFA should establish a greater diversity of contacts among these organizations. Where R&D efforts are proposed, the DDFA should sponsor such work only if a specific high-priority need has been identified. Technology development in this context should be assigned the role of addressing those needs that will remain after 2006. Sponsorship should be directed toward institutions having demonstrated capability and performance.

References

Alm, A. L. 1996. Environmental Management Ten-Year Plan Initiative. Presentation to the Board on Radioactive Waste Management, October 24, 1996. Washington, D.C.: U.S. Department of Energy.

Alm, A. L. 1997. Memorandum on Technology Deployment. Washington, D.C.: U.S. Department of Energy.

Bedick, R., et al. 1996. Record of the Decontamination and Decommissioning (D&D) Workshop: Defining Research Needs. DOE/METC-96. Morgantown, W. Va.: U.S. Department of Energy.

Berkey, E. 1997. Testimony before the Committee on Commerce Subcommittee on Oversight and Investigations, U.S. House of Representatives, by Edgar Berkey, Chairman of the Committee on Technology Development and Transfer, Environmental Management Advisory Board. Washington, D.C.: U.S. Department of Energy.

Boyd, G. 1997a. Correspondence from Gerald G. Boyd, Acting Deputy Assistant Secretary for Science and Technology, to Dr. Peter Myers, Chairman, Committee on Decontamination and Decommissioning. Washington, D.C.: U.S. Department of Energy.

Boyd, G. 1997b. Presentation by Gerald G. Boyd, Acting Deputy Assistant Secretary for Science and Technology, to the Committee on Decontamination and Decommissioning. December 10, 1997. Washington, D.C.: U.S. Department of Energy.

General Accounting Office (GAO). 1996. Energy Management: Technology Development Program Taking Action to Address Problems. Report to the Ranking Minority Member, Committee on Governmental Affairs, U.S. Senate. GAO/RCED-96-184. Washington, D.C.: General Accounting Office.

General Accounting Office (GAO). 1997. Cleanup Technology, DOE's Program to Develop New Technologies for Environmental Cleanup. Testimony by V. S. Rezendes, Director, Energy, Resources, and Science Issues, GAO, before the Subcommittee on Oversight and Investigations, Committee on Commerce, U.S. House of Representatives. GAO/T-RCED-97-161. Washington, D.C.: General Accounting Office.

Hart, P. 1997. Presentation by Paul Hart, DDFA Program Manager, DOE-FETC, to the Committee on Decontamination and Decommissioning. Irvine, Calif. December 10, 1997.

Hyde, J. 1997. Presentation by Jerry Hyde, D&D Program Manager, DOE-HQ, to the Committee on Decontamination and Decommissioning. Woods Hole, Mass. June 23, 1997.

Kessinger, M. D. 1997. Letter from M.D. Kessinger, FETC Program Manager, U.S. Army Corps of Engineers, to P. W. Hart, USDOE, FETC, Morgantown, W. Va., August 26, 1997.

Kessinger, M. D. and W. Greenwald. 1997. Presentations by M. D. Kessinger and W. Greenwald, representing the U.S. Army Corps of Engineers, to the Committee on Decontamination and Decommissioning. Woods Hole, Mass. June 1997.

National Research Council (NRC). 1996. Environmental Management Technology: Development Program at the Department of Energy, 1995 Review. Committee on Environmental Management Technologies. Washington, D.C.: National Academy Press.

National Research Council (NRC). 1997. Building an Effective Environmental Management Science Program, Final Assessment. Washington, D.C.: National Academy Press.

Strategic Laboratory Council (SLC). 1997. Overview Recommendations for Office of Science and Technology. Briefing material sent to Mr. Alvin Alm by Mr. Terry Surles. May 15, 1997. Washington, D.C.: U.S. Department of Energy.

U.S. Army Corps of Engineers. 1997. Large-Scale Demonstration Project—Cost Standardization Study, July 1997.

U.S. Department of Energy (DOE). 1995a. Estimating the Cold War Mortgage: The Baseline Environmental Management Report (BEMR). DOE/EM-0232. March 1995. Washington, D.C.: U.S. Department of Energy.

U.S. Department of Energy (DOE). 1995b. Five Year Plan (FY 1991-95). DOE/S-0070. Washington, D.C.: U.S. Department of Energy.

U.S. Department of Energy (DOE). 1995c. Decontamination and Decommissioning Focus Area Technology Summary. DOE/EM-0253. June 1995. Washington, D.C.: U.S. Department of Energy.

U.S. Department of Energy (DOE). 1996a. Results of the National Needs Assessment. July 24-25, 1996. Morgantown, W.Va.: U.S. Department of Energy.

U.S. Department of Energy (DOE). 1996b. Decontamination and Decommissioning Focus Area Technology Summary. DOE/EM-0300. August 1996. Washington, D.C.: U.S. Department of Energy.

U.S. Department of Energy (DOE). 1996c. The 1996 Baseline Environmental Management Report (BEMR), DOE/EM-0290. June 1996. Washington, D.C.: U.S. Department of Energy.

U.S. Department of Energy (DOE). 1996d. Science Program Description. 9 Dec. 1996 [last update]. Online. Available: http://www.em.doe.gov/science/sciprog.html.

U.S. Department of Energy (DOE). 1997a. Decontamination and Decommissioning Focus Area (Predecisional Draft OST Linkage Table). Washington, D.C.: U.S. Department of Energy.

U.S. Department of Energy (DOE). 1997b. D&D Focus Area Annual Report 1996. DOE/FETC-97/1041. Morgantown, W. Va.: U.S. Department of Energy.

U.S. Department of Energy (DOE). 1997c. Accelerating Cleanup: Focus on 2006. DOE/EM 0327. June 1997. Washington, D.C.: U.S. Department of Energy.

U.S. Department of Energy (DOE). 1997d. Decontamination and Decommissioning Focus Area Review Report. FETC. November 4-7, 1997. Morgantown, W. Va.: U.S. Department of Energy.

U.S. Department of Energy (DOE). 1997f. Federal Energy and Technology Center, U.S. Department of Energy, D&D Focus Area Fiscal Year 1998 Program Review Briefing Material and Reviewer Comment Forms, Books 1 and 2. November 4-6, 1997. Morgantown, W. Va.: U.S. Department of Energy.

U.S. Department of Energy (DOE). 1997g. Call for Technology Deployment Initiative Proposals, Idaho Operations Office. March 7, 1997. Idaho Falls, Id.: U.S. Department of Energy.

U.S. Department of Energy (DOE). 1998a. Deactivation and Decommissioning Focus Area, Quarterly Report—January 1998. Washington, D.C.: U.S. Department of Energy.

U.S. Department of Energy (DOE). 1998b. Accelerating Cleanup: Paths to Closure. DOE/EM-0342. February 1998. Washington, D.C.: U.S. Department of Energy.

APPENDIXES

Appendix

A

Statement of Task

A REVIEW OF DECONTAMINATION AND DECOMMISSIONING TECHNOLOGY DEVELOPMENT PROGRAMS AT DOE

Committee on Decontamination and Decommissioning
National Research Council
Board on Radioactive Waste Management

The Department of Energy has estimated that the cost of cleaning up the nation's atomic weapons complex will be hundreds of billions of dollars and will require decades to complete. The Department has therefore assigned to its OST the task of developing innovative technologies that will make the cleanup of the weapons complex, including D&D of complex facilities, less costly, quicker, and less hazardous to workers, the public, and the environment.

The NRC Committee on Decontamination and Decommissioning will review the approaches used by the DDFA of OST to assess their utility and effectiveness. To this end, the committee will study how the DDFA selects and evaluates alternative D&D technologies and the methods of seeking deployment of the selected technologies by those charged with the actual cleanup work. The Committee will review the Large-Scale Technology Demonstration Project processes, and may review other techniques employed by the OST to enhance utilization of innovative cleanup technologies as they apply to the DDFA.

The Committee will review the processes that the DDFA is using to compare the advantages anticipated from technology development programs within DOE to existing cleanup processes that are available commercially. The Committee also will evaluate applicability and effectiveness of technology development activities in the DDFA.

The Committee will issue a report at the end of the study.

APPENDIX

B

Summary of the Committee's Meetings

In conformance with its Statement of Task, the committee sought to examine the technology development activities of the DDFA of the DOE OST. The committee tried to determine whether the needs of the sites to be cleaned up were consistent with OST development activities and why the field sites did not make greater use of the cleanup technologies that were identified or developed with OST funding.

The committee obtained most of its information from presentations made to the committee at its meetings by representatives of DOE and its contractors. The committee was fortunate to hear from the Assistant Secretary for Technology Development, the Director of OST, the head of the DDFA, and OST staff. The committee heard presentations from essentially all principal university, labor union, industry, and technical society contractors to the DOE. The committee attended DDFA annual review sessions and special DOE topical meetings. There were several occasions, however, when last-minute cancellations of scheduled DOE presentations delayed the committee's progress, as did occasions when written confirmation was found to supersede earlier oral information.

Committee members visited the three active LSDP sites, interviewed site and contractor personnel and recorded their observations (see Appendixes C-F). Full committee meetings took place at DOE or NRC facilities in Washington, D.C.; Morgantown, West Virginia; Woods Hole, Massachusetts; and Irvine, California. For the most part, the committee relied on the expertise of its members and the NRC staff for understanding and evaluating the information presented to it. Although individual committee members had discussions with university and industry experts on specific topics, the committee found it unnecessary to employ consultants.

The following is a summary of full committee meetings held during the review period 1996-97:

1. The committee first met in September 1996 for a preliminary discussion and planning session in Washington, D.C.

2. On December 12-13, 1996, the committee met at the Beckman Center in Irvine, California. During an executive session, committee members heard presentations regarding the trip reports that had been undertaken to date. Open sessions were held during the two-day meeting with presentations from Jerry Hyde, DOE-HQ Program Manager for DDFA, who discussed the DDFA program, METC and the status of the LSDPs.

3. The committee met again on April 1-3, 1997, for the Mid-Year Review of the DDFA program at METC. The committee found that of the 18 papers listed on the agenda, 3 were withdrawn, 3 were presented by alternate speakers, and many of the speakers were not conversant with the subject they presented. After the presentations, the committee had an informal discussion with Paul Hart, DOE-METC Focus Area Lead, regarding the LSDPs. Following the Mid-Year Review, the committee met in closed session to discuss the presentations.

4. On June 23-24, 1997, the committee met at the NRC's Woods Hole Study Center. They reviewed the committee's work to date, produced a draft outline, and made writing assignments during the closed sessions. During the open sessions of this meeting, Mark Kessinger and Wendell Greenwald, representatives of the U.S. Army Corps of Engineers, discussed the Corps' work with the DDFA and gave an overview of its costing procedure and answered questions from the committee. Jerry Hyde discussed the budget of EM, OST, and its programs, and summarized the TDI—its mission, objectives, and budget plan. He went on to discuss the LSDPs, including an explanation of the D&D product lines, status of current demonstration projects, and future schedule. Mr. Hyde discussed in particular the interim safe storage LSDP at the Hanford C-Reactor. Concerns were voiced by the committee about the unavailability of the "Green Books." The representatives from the Corps participated in this discussion.

Later, Dave Clements, project manager for the BNFL Engineering, Ltd., D&D project at Capenhurst Diffusion Plant, England, gave a presentation on his experience and knowledge of D&D technologies in the field. Fred Petschauer of Brookhaven National Laboratory presented his experiences with the D&D of the Shoreham Nuclear Plant.

5. A meeting of the committee was held again in Washington, D.C., on September 25-26, 1997. The closed session included a presentation of the final site visit trip report, and discussion of new information provided to the committee

by Jerry Hyde, Paul Hart, and Steve Bossart, a DDFA Project Manager from FETC (nee METC). The remainder of the meeting was spent drafting the report.

6. At the December 10-12, 1997, meeting in Irvine, California, the open session included a video conference with Gerald Boyd, Acting Deputy Assistant Secretary for OST, who provided additional information OST regarding the DDFA and LSDPs. Jef Walker, DOE OST, and Paul Hart, FETC, were present with the committee. The afternoon was spent in discussions of the relationship between OST and the DDFA, and the LSDP's role, past and future, in technology transfer to DOE sites. The executive session included discussion of the video conference and initial preparation of the first draft report.

7. In March 1998, the committee again met in Washington, D.C., for a writing session to finalize a draft of the report.

C

Summary of the meeting held on April 18, 1996, at the Argonne National Laboratory on the subject of the DOE Large-Scale Technology Demonstration Program associated with the D&D of the Chicago Pile 5 (CP-5) Reactor

Trip Report by
Trish Baisden and Frank Crimi for the
NRC Subcommittee on Decontamination and Decommissioning
Submitted June 3, 1996

Attendees

Dick Baker	DOE Chicago Operations Office
Rob Rose	Project Manager, Technology Development Division, Decontamination and Decommissioning Projects, D&D Project Engineer for CP-5, ANL
David Black	Waste Management Engineering, Technology Development Division, ANL
Tom Yule	Manager, Waste Management Program, Technology Development Division, D&D Program Leader for CP-5, ANL
Jerry Hyde	DOE HQ, EM-50
Steve Bossart	METC, Morgantown, WV
Frank Crimi	Lockheed Environmental Systems, NRC Subcommittee on D&D
Trish Baisden	Lawrence Livermore National Laboratory, NRC Subcommittee on D&D

The meeting summary and response to questions represent our understanding gained during this visit with DOE and Argonne National Laboratory (ANL) personnel. Numbered questions in the following text (e.g., Question 1), were sent to METC prior to the ANL visit. Questions that arose during the visit are not numbered.

MEETING SUMMARY

Some history regarding the ANL proposal submission for funding associated with DOE's LSDP was provided. The METC representative explained that in July 1995, DOE-EM put out a Request for Proposals through the METC organization for the LSDP. Eight proposals were received; two were submitted by the Savannah River Laboratory, one from Oak Ridge National Laboratory, one from Rocky Flats, one from Battelle, one from Hanford for the C-Reactor, one from Fernald for its Plant 1 Facility, and one from ANL for the Chicago Pile-5 (CP-5) Reactor decommissioning. It was further explained that EM-40 (the Office of Environmental Restoration) and EM-60 (the Office of Facility Transition and Management) within DOE would be the organizations doing the testing, but EM-50 was the holder of technologies. The schedule for an LSDP was set by DOE to be 18 months and three technology projects were to be selected for the first year of the program. The second year, two more projects would be selected and in the third, three more projects. By the year 2000, DOE, through the DDFA, expects 90 percent of the technologies to have undergone a large-scale technology demonstration.

ANL commented that the DOE METC solicitation required that the proposal incorporate the Integrating Contractor Project Management concept (i.e., integrating industry, university, national laboratory, and international expertise to accelerate technology progress) and that the proposals be endorsed by and submitted through a DOE field operations[1] office. The latter requirement was to ensure commitment to the LSDP from the respective field operations offices.

In October 1995 METC awarded LSDP contracts to ANL for CP-5, to Hanford for the C-Reactor, and to Fernald for Plant 1.

DOE briefly described the use of cooperative agreements between the DOE Chicago Field Office (DOE-CH) and the Strategic Alliance as well as differences between commercial business practices and the Federal Acquisition Regulations (FAR). DOE also discussed three barriers or issues that had to be overcome or resolved before the CP-5 contract could be awarded:

1) Intellectual Property—At the present time, DOE appears to be giving up rights to the property and patents as an incentive to companies to test their technologies;

---

[1]Proposal for a Large Scale Decontamination and Decommissioning (D&D) Demonstration at the CP-5 Facility, August 18, 1995, Chicago Operations Office, Department of Energy.

2) Product Liability—The liability limit is $25 million with the Strategic Alliance liable for the first $15 million; DOE pays all claims above this amount; and,

3) Issues related to negligence.

QUESTION 1: What led DOE-CH to propose its site for a D&D LSDP? For example, was there one particular problem you needed help on or did you include the LSDP so you could stretch your D&D budget?

ANL responded that ANL is the best place to develop technology for D&D because of ANL's commercial, naval, and research reactor experience. ANL has the right people with the corporate memory. ANL has experience in the D&D arena already. It is not one particular problem we are looking for help on but rather ANL is committed to developing and testing new technologies. D&D of nuclear facilities is one of ANL's strategic thrusts and members of the staff have been working for over two years developing an overall D&D initiative. When the METC call for proposals was made, we were prepared to respond.

The proposal submitted through DOE-CH requested a total of $5.25 million from DOE-HQ EM-50 for the period of October 1995 through March 1997 to accommodate and demonstrate approximately 40 technologies. This request included anticipated costs for technology evaluations, interface engineering, safety assessments, and demonstration costs. It was proposed that after finalizing the selection of technologies the budget allocations for the specific LSDPs be prepared jointly by DOE-CH and METC staff.[2]

ANL also commented that organizational changes were made to facilitate its D&D initiative. At ANL, the D&D effort was moved from the Operations Division to the Technology Development Division.

Within the ANL management structure, all project managers are members of the Technology Development Division.[3] ANL is attempting to marry users of technology with developers of technology. ANL managers are required to identify and develop new technologies and be willing to take a risk.

ANL has personnel with extensive D&D experience. For example, the ANL CP-5 D&D Project Engineer who was present at the meeting was the Project Manager for the recently completed D&D efforts on both the B-200 M-Wing hot cells and the B-212 D-Wing Plutonium Gloveboxes at ANL. Another example given was an individual who was the Project Manager for the recently completed D&D of the Experimental Boiling Water Reactor (EBWR) and is now the Project Manager for the D&D of the JANUS reactor.

ANL also mentioned that some of the funds originally earmarked for its Integral Fast Reactor (IFR) Program are being redirected into a couple of decon-

---

[2]Ibid., Section II.B

[3]Organizational chart for the Technology Development Division, the organization unit within ANL involved in the CP-5 D&D LSDP.

tamination projects at ANL, one based on a laser and the other on chemical decontamination technologies.

Since use of the Integrating Contractor concept was a METC requirement for proposals, ANL was asked on what basis they chose their partners (i.e., members of the Strategic Alliance).

DOE-CH established the Strategic Alliance with ANL and the others in response to the condition in the Request for Proposals that the integrating contractor management concept be used for the proposed LSDP of D&D technology. DOE-CH is committed to co-manage the LSDP and the D&D integrating contractor with METC.

Members of the Strategic Alliance for the LSDP at CP-5 include:

Partner	*Reasons Selected*
Duke Engineering & Services	Commercial nuclear experience and demonstrated cost effective management practices
ICF–Kaiser	Experience with commercializing technologies, particularly in the international marketplace
Florida International University (FIU)	17 universities were evaluated; chose FIU because of its expertise in decontaminating concrete
3M	3M has several technologies that may be cost-effective alternates to baseline technologies of the CP-5 reactor demonstration
Commonwealth Edison (ComEd)	Interface with commercial utilities
Argonne National Laboratory	Site of facility to undergo D&D

In developing a list of potential partners for the Strategic Alliance, did DOE issue a *Commerce Business Daily* (CBD) announcement for an expression of interest concerning the decommissioning of CP-5?

The reply from DOE–Chicago was that no CBD notice was issued for CP-5.

How are the D&D efforts being managed by the Strategic Alliance?

ANL responded by saying they are using performance-based management rather than compliance-based management. ANL explained that all of the technology providers (TP) share in the risk. Their cost share is 30 percent of the total cost (i.e., a fixed-price contract) for the development and demonstration of a LSDP project. Members of the Strategic Alliance who also want to participate as technology providers were required to declare their intentions at the time the alliance was being formed. 3M and FIU made this declaration. Under the conditions of the Strategic Alliance, ANL does not have to declare and can be both a member of the Strategic Alliance and a technology provider.

Articles of Collaboration, an agreement among the members of the Strategic Alliance that defines how the members will interact and resolve disagreements, were also developed. The Articles of Collaboration provide for the election of a Board of Directors. ANL is a signatory to the Articles of Collaboration. DOE sits on the Board of Directors as a non-voting member. The Board of Directors works on a 90-day window for reviewing and approving projects. Once the Board of Directors makes it decision, then it goes to DOE-CH, METC, and EM-40 for approval.

Technology providers often have good candidate technologies, but they do not have experience working in a radiation environment. ANL expects the D&D of CP-5 to be a good place to learn how to work in a radiation area because of the relatively low radiation levels associated with CP-5. However, the increased cost associated with training workers and working in a radioactively contaminated area will be representative of future applications of D&D technology.

Members of the companies that serve on the Board of Directors cannot participate in the demonstration of one of their technologies and simultaneously serve on the Board. If they want to demonstrate one of their technologies, they become a non-voting member while their technology goes through the selection process and, if selected, during the time the technology is evaluated for overall effectiveness and performance. If, for example, 3M, which is on the Board of Directors, wants also to act as a TP, it cannot vote on the performance of its technology. 3M must step out of the Strategic Alliance when it is demonstrating its technology.

The representative from DOE-CH stated that the Board of Directors was expected to be a self-policing group. It is to establish performance criteria and metrics by which to judge performance (i.e., 90-day performance milestones). It was also commented that while the first million dollars was being expended, DOE-CH would stand back and let the Strategic Alliance do its job. The report card at the end will tell the story of how efficiently and cost effectively the Strategic Alliance has both managed the D&D efforts to decommission CP-5 and integrated the testing of new D&D technologies.

The functions and the organizational structure of the DOE-CH D&D LSDP are shown in Figure 1.

How are the D&D funds managed?

A Cooperative Agreement was drawn up between members of the Strategic Alliance (with the exception of ANL). The Cooperative Agreement was devel-

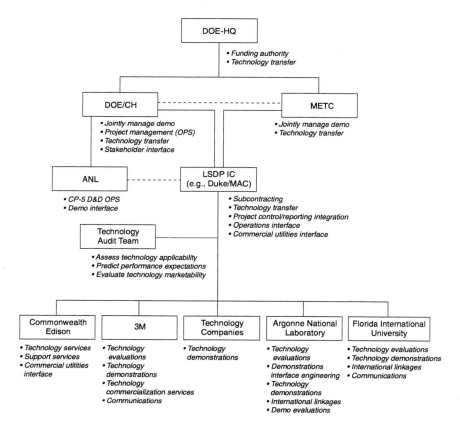

FIGURE 1   Functions and Organization Structure of the DOE-CH D&D LSDP.

oped to cover the cost share commitment associated with the Technology Providers, as well as issues related to liability and intellectual property. A Cooperative Agreement is being used that incorporates commercial business practices rather than the FAR requirements. ANL is not a signatory to the Cooperative Agreement because liability and intellectual property rights are covered for them under the contract between DOE and the University of Chicago. It was reiterated that the biggest issue related to developing documents such as a Cooperative Agreement was the question of liability. Intellectual property was also a stumbling block, but DOE gave up property rights and patents as an incentive.

After the LSDP contract was awarded to ANL by METC in October 1995, ANL entered into a Basic Ordering Agreement (BOA) with Duke Engineering & Services so that the planning process could begin. The BOA offered ANL the greatest flexibility in that the agreement could be amended quite easily to increase the scope of work. As new, well-defined tasks are added, Duke Engineering & Services provides ANL with a revised cost estimate to cover the increased scope of work. The revised cost is based on a previously agreed upon rate structure for services. Duke Engineering & Services is handling all of the administrative details for the Strategic Alliance and will be tasked with writing all the contractual agreements with technology providers for the LSDPs.

In early 1996 METC asked for a separate proposal to be submitted for the LSDP for CP-5, but this time the proposal was to be from the Strategic Alliance on behalf of ANL and not from DOE-CH. METC also asked for the proposed work to be submitted by the Strategic Alliance using the Technical Task Proposal (TTP) format required by DOE-EM. The Strategic Alliance responded to both of these requests. The new proposal and the associated TTP is the mechanism that allows the DOE to fund members of the Strategic Alliance directly.

How are LSDP technologies selected and if selected, how are they evaluated?

Technologies are selected by a Technology Selection Team (TST). Members of the TST for the LSDP for CP-5 include:

Organization	*Representative*
ANL	David Black
ComEd	Rock Akers
Duke Engineering & Services	Brian Cruse
FIU	Joe Boudreaux
ICF–Kaiser	Holmer Duggar (Chair)
3M	Jack Everett
METC (ex-officio)	Steve Bossart

Functions of the TST were described as:

- Identifying needs and developing problem sets;
- Developing selection criteria;
- Looking for innovative demonstration technologies;
- Evaluating technologies;
- Recommending technologies for demonstration;
- Specifying performance indicators for demonstration; and
- Assessing and reporting on technologies demonstrated.

The Screening/Selection Criteria developed by the TST include:

- State of maturity;
- Improvement over baseline;
- Application across complex/cost benefit;
- Transportability to CP-5;

- Cost of CP-5 demonstration;
- TP interest;
- Compatibility with CP-5 schedules; and
- Value of information gained by demonstration.

After initial selection, other facility specific issues of the proposed technology demonstration are established. These include:

- Does the LSDP fall within the established safety and environmental envelope?
- Are there any issues related to characterization and waste disposal?
- Can the utility/support system requirements reasonably be met?

Table 1 lists the applicability of EM-50/METC funded technologies to the CP-5 LSDP. The members of TST evaluated primarily METC/EM-50 supported technologies. Individual members of the TST are charged with evaluating technologies in their own areas of expertise. Table 2 lists the potential technology demonstration opportunities in CP-5 by activity.

Factors considered in technology evaluation:

- Improvements over baseline;
- Reduction of secondary waste volume;
- Cost;
- Personnel exposure;
- Time;
- Training, viability; and
- Job specific issues.

Technology demonstration will be done in two phases:

1. Test plan development and safety analysis (may stop here because of facility specific issues); and
2. Demonstration performance (following approval of the initial plan with a "90-day window" to allow for changes in the plan).

At the end of the demonstration, a technical report will be generated by the Test Engineer. The report will contain "just the facts" and no conclusions. The TST will use the technical report and other information to make a final evaluation and provide conclusions and recommendations. This information will be documented in a final report issued by the TST.

How does ANL make cost estimates for D&D activities?

ANL has worked closely with Nuclear Engineering Services (NES), a private company specializing in decommissioning consulting and field support with experience in D&D planning, cost estimating, remediation, field services, and radiological health and safety, to develop the cost estimates. In essence, NES is validating ANL's cost estimates. ANL commented that good characterization data is needed before going out to bid. Otherwise, the scope of the job can change drastically. Issuing change orders causes the costs to escalate.

ANL stated that it wants to develop the capability to do good cost estimates internally. For benchmarking cost estimates, DOE-CH Operations Office used TLG Services, Inc., a company similar to NES, to validate cost estimates independently. ANL also believes that it gained some valuable experience in this area with the D&D of the EBWR.

ANL plans to use an internally bid contract (i.e., use ANL personnel) for health physics support of D&D efforts at ANL and for characterization of the JANUS reactor.

In developing the plans for the D&D of the CP-5 reactor, did ANL rely on documentation associated with the D&D of the Shippingport reactor?

DOE-CH answered that it did use the Shippingport documentation in planning CP-5.

QUESTION 2: What baseline technologies were being considered prior to the insertion of the LSDP? What would you use if there were no mature innovative technologies?

The Strategic Alliance selected baseline technologies on the basis of comfort, known technologies that will work, and their collective experience. A baseline technology or short list of needs (if no baseline technology is available) was selected for each D&D operation associated with the decommissioning of CP-5. All baseline technologies are documented in the overall CP-5 Decontamination and Decommissioning Plan. (Because this information is contained in several large reports, copies were not requested.)

How do you prevent getting involved with an immature technology, that is, a technology which is not sufficiently developed to be tested as an LSDP?

The reply was that immature technologies should be weeded out by the provision requiring the technology provider to cost share at the 30 percent level. DOE does not want to pay for technology development if it is not ready for a

TABLE 1 Applicability of RM-50/METC funded technologies to CP-5 LSDP

EM-50/METC FUNDED TECHNOLOGIES

D&D Focus Area Technology Summary	Lower life-cycle costs	Lower H&S risks to worker and public	Lower risks for detrimental impact to environment	Reduced quantity of waste materials for disposal	Reduced amount of secondary waste	Reduced hazard level and category of waste	Increased material reuse and/or free release	Less time required for char/deact/D&D/dispos	Reduced amount of residual contamination	Reduced cost and worker exposure in S&M	Overall Ranking (to be determined)	Comments
Protective Clothing - Permselective Membrane and Carbon Adsorp.	L	M	NA	L	L	NA	NA	M	NA	M		
Advanced Worker Protection System	M	H	NA	M	M	NA	NA	H	NA	M		
Characterization of Rad. Contamination in Pipes with Pipe Explorer	M	M	NA	M	L	NA	NA	M	NA	M		
Three-Dimensional Integrated Characterization and Archiving System	M	M	NA	M	M	NA	M	M	NA	H		
Decontamination and Recycling of Concrete	M	M	L	L	L	M	L	NA	M	NA		
Remote Operated Vehicle Dry Ice Pellet Decontamination System	L	M	M	M	H	L	L	M	M	NA		
Laser Ablation of Contaminants from Concrete and Metal Surfaces	L	M	M	M	M	L	M	L	M	NA		
Chemical Decon. of Process Equip. Using Recyclable Chelate												To be determined

H = High,   M = Medium,   L = Low Probability of Significant Benefit

Technology										K-25 Site Demo
Soda Blasting Decontamination Process										
Decontamination of Gaseous Diffusion Process Equipment										Not applicable
Concrete Decontamination Using Microwaves	L	M	M	M	M	L	M	M	NA	
Concrete Decontamination by Electro-Hydraulic Scabbling	L	M	M	M	M	L	M	M	NA	
Electrokinetic Decontamination of Concrete	L	M	M	H	M	L	M	M	NA	
Decontamination and Conversion of Nickel Radioactive Scrap Metal										Not applicable
Mobile Work System for Decontamination and Decommissioning	M	H	NA	M	NA	NA	M	NA	M	
Asbestos Pipe-Insulation Removal System										To be determined
Stainless Steel Beneficial Reuse Demonstration										To be determined
Recycle of Contaminated Metals										To be determined
Technology Development Data Sheets										
Coherent Laser Vision System										Not applicable
Rapid Surface Sampling and Archive Record (RSSAR) System										Not applicable
Advanced Technologies for Decon. and Conversion of Scrap Metal										To be determined
Interactive Computer-Enhanced Remote Viewing System										Not applicable
Robotic Manipulator Monitoring and Diagnostics										Not applicable
Catalytic Extraction Processing of Contaminated Scrap Metal										Not applicable
Portable Sensor for Hazardous Waste	M	M	NA	NA	L	L	NA	H	H	
Reconfigurable In-Tank Mobile Robot										Not applicable
Treatability Study Using Prompt Gamma Neutron Activation										To be determined

TABLE 1 Continued

EM-50/METC FUNDED TECHNOLOGIES (cont.)

H = High, M = Medium, L = Low Probability of Significant Benefit

Technical Task Plans (TTPs)	Lower life-cycle costs	Lower H & S risks to worker and public	Lower risks for detrimental impact to environment	Reduced quantity of waste materials for disposal	Reduced amount of secondary waste	Reduced hazard level and category of waste	Increased material reuse and/or free release	Less time required for char/deact/D&D/dispos	Reduced amount of residual contamination	Reduced cost and worker exposure in S&M	Overall Ranking (to be determined)	Comments
Equipment for Remote Disassembly of Highly Radioactive Equipment												To be determined
Remote Removal, Handling, and Size Reduction of Large Equipment												To be determined
Standardized Dismantlement System for Reactor Decommissioning												Not funded
Remote Laser Processing of Radioactive Components												T-Plant Demo
Improved Concrete Cutting Methods												Paper Study
Massive Equipment Size Reduction and Removal System	M	H	NA	M	M	NA	NA	M	NA	M		
Tooling System for Massive Equipment Size Reduction and Removal												To be determined
Accelerated Testing of Concrete Decontamination Methods												Demo Program

Technology											Status
Microwave Concrete Decontamination System	L	M	M	L	M	M	L	M	M	NA	
Remote Mechanical Methods for Decontaminating Concrete Walls											To be determined
Decontamination Using Liquid Nitrogen Carrier with Solid CO_2	L	M	M	L	H	M	L	M	M	M	
Long-Term, Low-Temperature In-Situ Decontamination Demonstration											Not applicable
In-Situ Flushing Decontamination in Equipment Interiors											Not applicable
Liquid Handling for Plutonium Facilities											Not applicable
Stabilization of Reactor Fuel Storage Pool											Not applicable
Portable Concentrator for Processing Pu Contaminated Solutions											Not applicable
Demonstration of Selective Size Reduction on Plutonium Glovebox											Not applicable
Plutonium Glovebox Disposition System											Not applicable
Conversion of Asbestos Material into a Non-Regulated Material											To be determined
Mineralogical Conversion of Asbestos Waste Technology											To be determined
Depleted Uranium Recycle Program											Not applicable
Stainless Steel Beneficial Reuse Demonstration											To be determined
Metal Recycle Technology Development											To be determined
D&D Robotics											
Swing Free Crane Control	M	H	NA	NA	NA	NA	M	NA	NA		
Mapping, Characterization, and Inspection System (MACS)	M	H	NA	M	M	M	M	NA	H		
Selective Equipment Removal System (Dual-Arm Work Module)	M	H	NA	NA	M	NA	M	NA	NA		
Internal Duct Characterization System	M	H	NA	M	M	M	M	NA	NA		

TABLE 2   Potential Technology Demonstration Opportunities in CP-5

Activity	Demonstration Location	Possible Technologies
1. Characterization	Rod storage holes Internal piping surfaces Exhaust system ductwork Automated surveys (floors, wall)	a. Characterization of Radioactive Contamination Inside Pipes with the Pipe Explore™ System b. Internal Duct Characterization System c. Three-Dimensional Integrated Characterization and Archiving System (3D-ICAS) d. Mapping, Characterization, and Inspection System (MACS-Mobile Automated Characterization System)
2. Concrete Decontamination	Concrete floors, walls Wall storage area Hot cell	a. Decontamination Using Liquid Nitrogen Carrier with Solid CO_2 Particles b. Remote Operated Vehicle Dry Ice Pellet Decontamination System c. Laser Ablation of Contaminants from Concrete and Metal Surfaces d. Concrete Decontamination Using Microwaves e. Electrokinetic Decontamination of Concrete
3. Remote Dismantlement	Reactor vessel Graphite reflector Thermal shield Lower shield Biological shield	a. Dual-Arm Work Module b. Mobile Work System for Decontamination and Decommissioning (ROSIE) c. Swing-Free Crane Control
4. Worker Protection	In-cell work	a. Advanced Worker Projection System b. Protective Clothing Based on Permselective Membrane and Carbon Adsorption
5. Work Area Containment	Reactor block	a. Innovative Containment Design b. Capture Ventilation Systems c. Contained Waste Transfer Equipment

LSDP. Contracts will be negotiated for a fixed price. The cost share requirement applies only if the contractor is testing a technology.

QUESTION 3: What was your comfort level with the baseline technologies? What is your comfort level with the proposed LSDP technologies?

ANL feels comfortable with the technologies that already have undergone the selection process. The technologies that are ready to go include:

- Mobile Automated Characterization System (MACS)—ORNL-developed technology for automated floor surveys for radiation contamination (expected early May 1996);
- Membrane Filtration—3M-developed technology for the removal of Cs-137, Co-60 from pool water (Gary Goken, 3M, is the contact);
- Pipe Explorer™—Science and Engineering Associates-developed technology for characterization of contamination in underground drain and ventilation piping (expected late April 1996);
- Rotopeen—3M-developed technology for the removal of fixed contamination from rod storage hole liners. This is a scabbling technique using tungsten-carbide blades, but there is an issue involving use of the Rotopeen in a radiation environment. The baseline technology for this operation is cutting the pipes out of concrete.

Future demonstration technologies that have been selected include:

- Contaminated paint removal from concrete:
  ∘ Flashlamp—high intensity light for heating painted surfaces
  ∘ ROVCO$_2$—Remote Operation Vehicle with solid CO$_2$ blasting
  ∘ Rotopeen
  (ANL has already tried some chemical decontamination and grit blasting)
- Advanced Worker Protection System
- Robotics
  ∘ Dual Arm Work Module
  ∘ ROSIE-II

Why are you using robotics when a job can be done by hand?

Previously the DOE EM-40 Program Manager stated that using robotics will drive up the costs especially if you use robotics where you can do the work by hand. Work can usually be done quicker by hand. Using robotics will take longer and increase costs. However, if using robotics can make a significant difference in areas where hands-on work is unsafe due to radiation levels, DOE is interested in developing it. For robotics to be a successful development effort, it must be tested in areas where hands-on work also can be done. If problems arise with the robotics during testing, workers need to be able to make repairs and adjustments without encountering unacceptable exposures to radiation.

What is the status of LSDP activities that are now on the CP-5 D&D schedule?

None of LSDP technologies so far selected are on a critical path of the CP-5 D&D schedule. ANL says it has some flexibility with EM-40. For example, the

D&D of the fuel pool is scheduled for 1998. However, time may permit the insertion of the Membrane Filtration LSDP ahead of schedule. ANL may ask for an increment of funding from EM-40 to do the membrane filtration test this year rather than waiting until 1998.

QUESTION 4: Are there contingency plans if the LSDP is unsuccessful? How long are they (i.e., ANL in concert with DOE EM-40) willing to wait in terms of delays to accommodate problems associated with the LSDP technology demonstrations?

ANL provided a copy of its project schedule which shows technology insertion points. The schedule has an outlook at of at least 180 days. Duke Engineering will write contracts that will have a termination clause if technologies are not working out. Technology providers will not be paid until the negotiated milestones are met. Contracts are written for service costs plus a fixed price for meeting the milestones. Work-arounds are laid out for contingency planning purposes. The schedule[4] contains parallel paths for LSDPs and are laid out so they are not all on the critical path.

Under which set of Environmental Safety and Health (ES&H) regulations do the D&D activities of CP-5 fall?

D&D activities funded by EM-40 have been going on at CP-5 for about two years and they have been operating under the usual ES&H regulations, primarily those in 10CFR835. In 1994, as part of the planning for the D&D of the CP-5 reactor, the associated hazards were identified, defined, assessed, and documented in a Safety Analysis Review (SAR). More recently, another SAR was prepared that included the LSDPs associated with CP-5.

During the past year, ANL also participated in a pilot study for the DOE ES&H Necessary and Sufficient Standards (N&SS) Program. The objective of the N&SS pilot study with ANL was to review the hazards associated with a real D&D situation and based on the assessment of the hazards, make judgments as to which DOE orders and regulations should apply; that is which ones are necessary and sufficient. The N&SS deemed appropriate for the D&D of CP-5 were documented in the pilot study. This document is in the final approval phase at DOE headquarters. Organizations in the approval chain for the N&SS for CP-5 include the University of Chicago (representing ANL) and DOE-EH-1. Once the N&SS are approved, it is intended that it is this set of ES&H regulations that will be followed. Thus, the Strategic Alliance would be bound to abide by the N&SS document, if approved. (A copy of this document was not yet available.)

---

[4] EM-40/EM-50-METC CP-5 D&D LSDP Integrated Projects Baseline Schedule.

A comment from one of the ANL staff was that many of the DOE orders can be traced back to a federal mandate.

DOE-CH pointed out that the D&D of CP-5 is not regulated under CERCLA (Comprehensive Environmental Response, Compensation, and Liability Act, "Superfund," enacted in 1980 and amended in 1986 as the Superfund Amendments and Reauthorization Act [SARA]). The State of Illinois supported a request for exemption from CERCLA. CERCLA/SARA pertains only to hazardous waste that is regulated by the Environmental Protection Agency and was intended to address the restoration of major U.S. uncontrolled hazardous waste sites, including landfills (Superfund sites). Since the D&D of CP-5 no doubt will involve the generation of radiological-only and mixed (radiological and hazardous) waste, ANL argued that its work falls more naturally solely under the Resource Conservation and Recovery Act (RCRA) (RCRA legislation was designed to regulate the production, handling, disposal, and reporting of hazardous wastes).

APPENDIX

D

Summary of the meeting held on November 7-8, 1996, at the Hanford Site on the subject of the DOE Large-Scale Technology Demonstration Program associated with the D&D of the 105 C-Reactor

Trip Report by
Trish Baisden and Frank Crimi for the
NRC Subcommittee on Decontamination and Decommissioning (D&D)
Submitted December 3, 1996

Attendees

Shannon Saget	DOE EM-50 Project Manager, Hanford Site
Jeff Bruggeman	DOE EM-40 Project Manager, Hanford Site
Greg Eidam	105 C-Reactor D&D Project Manager, Bechtel Hanford, Inc.
Jerry Hyde	DOE HQ, EM-50
Frank Crimi	Lockheed Environmental Systems, NRC Subcommittee on D&D
Trish Baisden	Lawrence Livermore National Laboratory, NRC Subcommittee on D&D

BACKGROUND INFORMATION

For over 40 years, the Hanford Site manufactured nuclear materials for the nation's defense programs. To assist in this nuclear materials production, nine water-cooled graphite-moderated plutonium reactors were constructed along the Columbia River by the U.S. government between the years 1943 and 1971. Eight of these reactors (B, C, D, DR, F, H, KE, and KW), operated between 1944 and

58

1971, have been retired from service and declared surplus. In 1989, the more general need to remediate radioactive and chemical waste at the Hanford Site was identified by the DOE, the EPA, and the Washington State Department of Ecology in the Hanford Federal Facilities Agreement and Consent Order, commonly know as the Tri-Party Agreement. In 1992, the need to remediate (D&D) eight of the surplus production reactors to protect the environment was identified in the Hanford Environmental Impact Statement (EIS). The following year, the DOE issued a Record of Decision (ROD) under the signature of Thomas Grumbly, then Assistant Secretary for Environmental Restoration and Waste Management, concerning the decommissioning of the eight reactors. The ninth reactor, N-Reactor, was in transition to deactivation and was therefore not considered within the scope of the EIS or the ROD.

In the ROD five decommissioning options were evaluated. These options were:

1. Safe storage followed by deferred one-piece removal that would include a safe storage period during which surveillance, monitoring, and maintenance would be continued until each reactor block could be moved intact, via a tractor-transporter, from its present location in the 100 Area to the 200 West Area for disposal (a distance of about 5-14 miles depending on the reactor location relative to the disposal site). A hypothetical safe storage period of 75 years was used to estimate the radiological inventory for this option.

2. No action; continue the present routine surveillance, monitoring, and maintenance of reactor structures for an indefinite period.

3. Immediate one-piece removal of the reactor block from its present location in the 100 Area to the 200 West Area for disposal.

4. Safe storage followed by deferred dismantlement and disposal at the 200 West Area.

5. In-situ decommissioning involving removal of roofs, superstructures, and concrete shield walls down to the level of the top of the reactor block and covering the remaining structures to a depth of at least 5 meters with earth and gravel. The resulting mound would be topped with an engineered barrier designed to limit water infiltration to 0.1 centimeter per year.

On the basis of a review of environmental impacts, total project cost, and the results of the public review process, the preferred option selected in the ROD was safe storage of the eight reactors followed by deferred one-piece removal.

WHY C-REACTOR?

C-Reactor was selected as the first reactor to be put into safe storage for several reasons. When start up of C-Reactor occurred in November 1952, C-Reactor was the largest reactor at that time, and consequently it became the principal test facility for exploring effects such as power level increases, graphite burn-out, and new fuel designs. Given the nature of testing, this reactor experienced higher-than-average fuel and process tube failures. In addition to the condition of the reactor core, a driver for selecting C-Reactor was high costs associated with maintenance and surveillance (M&S). With the exception of the D&D of some ancillary structures, stabilization of the fuel storage basins, and removal of the fuel, the reactor has been kept in an M&S mode since it was shut down in the spring of 1969. Age and resulting deterioration has made M&S an increasing expensive operation (about $1 million per year). Therefore, by demolishing as much of the building as possible and leaving only the structure and reactor core within the 3- to 5-foot thick concrete shielding wall, M&S and the potential for the spread of contamination can be reduced significantly. The other reason for choosing C-Reactor as the first of eight reactors to be decommissioned, was its proximity (900 feet south) to B-Reactor. Because B-Reactor has significant historical value, it has been nominated for inclusion in the National Register of Historic Places and currently is listed in the National Register by the National Park Service. Further, B-Reactor has been proposed as a publicly accessible museum, which carries with it certain implications as to the overall desired conditions at the combined 100 B/C Area. The last reasons for selecting C-Reactor were that the EIS ROD for the proposed D&D approach had been finalized, and the conceptual design and detailed radiological and hazardous materials characterization had been completed to implement the plan.

Note: Although Fluor-Daniel is the Site Management and Integration Contractor (I/C), all of the surplus reactors were turned over to Bechtel Hanford, Inc. (BHI), the Hanford Environmental Restoration Contractor.

DEFINITION OF COCOONING

"Cocooning" (i.e., safe storage) would involve the removal of 105 C-Reactor Building equipment and structure, except for the 3- to 5-foot concrete shielding walls adjacent to all sides of the reactor. These remaining shielding walls will be extended with concrete to the level of the highest shielding wall, and a roof system will be installed. An entry opening for access to the reactor area will remain, but it will be modified to accept a security door. C-Reactor ancillary buildings and structures would also be removed, leaving only the reduced reactor block enclosure.

105 C-REACTOR CONCEPTUAL DESIGN AND D&D PLAN

On September 29, 1995, BHI awarded a contract to Delphinus Engineering to develop a conceptual design for the interim safe storage of C-Reactor. The working definition of the safe storage condition given to Delphinus was:

- Interim inspection can be limited to once every five years.
- Containment to ensure that releases to the environment are not credible under normal design basis conditions.
- Shield wall penetration closures will maintain air/water leak tightness; in addition, below-grade penetration seals will withstand soil pressure and surcharge pressure.
- New roofing to be adequate to eliminate the need to replace the roof during the intended facility safe-store lifetime of 50 years.
- The final safe storage configuration of the facility not to preclude or significantly increase the cost of any decommissioning alternative for the reactor assembly itself.

The major activities addressed by Delphinus Engineering in its Conceptual Design Report (CDR) included:

- Site preparation, including security;
- Procurement strategy for long-lead-time equipment, materials, and services;
- Decontamination activities, including hazardous materials and equipment removal, building and structure demolition, and disposition of the Fuel Storage Basin and Transfer Bay;
- Waste management, including volume reduction, packaging, transportation, and disposal;
- Safe storage enclosure system design, including roof, penetration seals, electrical and remote monitoring, and ventilation; and
- Site restoration.

In April 1996 Delphinus Engineering delivered to BHI the CDR, which included the following deliverables:

- Conceptual design;
- Engineering studies;
- Long-lead-time procurement items;
- Preliminary hazards analysis;
- Work breakdown structure and WBS dictionary; and
- Conceptual cost estimate and project schedule.

Of particular note are the engineering studies. Five separate studies were performed to address the following areas:

- Safe storage enclosure;
- Decontamination versus contaminant removal;
- Demolition techniques selection;
- Salvage versus scrap value; and
- Disposition of the Fuel Storage Basin and Transfer Bay.

The purpose of the engineering studies was to conduct an in-depth search, determine potential alternatives, evaluate and compare alternatives, and develop prudent and cost-effective directions for the major design aspects of the projects.

The CDR developed by Delphinus Engineering became the basis for the 12-volume D&D plan from April through September 1996. The D&D plan was entitled "Definitive Design Report for the Reactor 105-C Interim Safe Storage Project."

RESULTS OF HANFORD SITE VISIT

The visit included a tour of the 105 C-Reactor Site on November 7, 1996, and discussions at the EESB Building in the 300 Area of the Hanford Site on November 8, 1996. In preparation for the site visit, a memorandum was sent to Jef Walker, DOE, Washington, and Shannon Saget, DOE, Hanford, by K. T. Thomas (dated October 18, 1996) on behalf of the NRC Subcommittee on D&D. The memo contained four questions to be addressed during the Hanford Site visit. These four questions (numbered), questions that arose during the visit, and our understanding of the BHI and DOE answers resulting from the discussions are given below.

Question 1: What led DOE-Richland (DOE-RL) to propose C-Reactor for a D&D large-scale technology demonstration (LSDP)? For example, was there one particular problem DOE needed help on, or did DOE-RL include the LSDP so as to stretch the D&D budget?

C-Reactor is one of eight reactors slated for safe storage followed by deferred one-piece removal. Developing new, practical technologies in the areas of characterization, decontamination, demolition, waste disposal, facility stabilization and worker health and safety during the C-Reactor project hopefully will reduce D&D costs, reduce health and safety risks, and expedite the safe storage of the other surplus reactors. The D&D of C-Reactor is being funded by EM-40 and EM-50 on a 50-50 cost-shared basis (projected total cost, $18 million.)

Question 2: What baseline technologies were being considered prior to the insertion of LSDP? What technologies would be used if there were no mature innovative technologies?

Baseline technologies are provided in the D&D plan for various work activities and in the CDR by Delphinus Engineering. As the design contractor, Delphinus Engineering was responsible for making a catalog of available technologies. In the CDR engineering studies, potential technologies were evaluated for the various activities identified (e.g., decontamination versus contaminant removal, demolition techniques, etc.). A description of each technique considered was included along with its potential application, applicability, cost, and availability. Technical references and source documents were included in each study. In determining baseline technologies, use was made of the DOE D&D Handbook (DOE/EM-0142P), and it is referenced repeatedly in the engineering studies.

Question 3: What is the comfort level with the baseline technologies? What is the comfort level with the proposed technologies?

The overall process used to select, review, and implement a LSDP associated with C-Reactor D&D is given in flowchart form in Figure 1. In the flow chart the roles of the IC Team and the Administrating Contractor are defined and the process used to determine, recommend, and implement the demonstration of innovative technologies is described. At this stage, care is taken to ensure that innovative technology has a direct comparison to the baseline technology.

Question 4: What are the contingency plans should the LSDP be unsuccessful? How long is DOE-RL willing to wait in terms of delays to accommodate problems associated with technology demonstrations for the LSDPs?

Environmental technology demonstrations are scheduled in parallel with the critical path activities and are not allowed to impact the schedule. The C-Reactor Safe Storage project is slated to be completed in about 18 months. Only mature technologies will be accepted for the C-Reactor LSDP. If a technology demonstration does not perform within the allotted time, then the original baseline technology will be used.

Is DOE-RL using an I/C for its LSDP? Was the I/C concept a METC requirement for proposals? What basis was used to choose the I/C partners (i.e., did they have to be members of the Strategic Alliance?)?

The use of an I/C was a requirement of METC for all LSDPs. According to the original proposal, DOE-RL is committed to managing the project coopera-

64

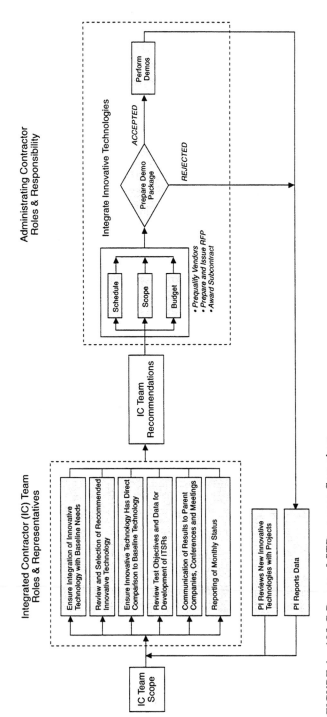

FIGURE 1   105-C Reactor LSTD process flow chart.

tively with METC. The Assistant Manager for Environmental Restoration (AME) will be the senior DOE-RL manager with direct responsibility for the C-Reactor demonstration project. BHI is the Environmental Restoration Contractor and will function as the I/C. BHI reports directly to the AME for all work performed under the Site ER contract. The C-Reactor LSDP is using an I/C Team approach which couples a Technology Alliance Team (TAT) with BHI Project Management, which serves as the Administrative Contractor. The TAT consists of representation from CH₂MHiLL, International Technology, Thermo Electron, Savannah River Site, Bechtel National, Inc., AEA Technologies, D&D subgroups, Montgomery-Watson, and Morrison Knudson. These organizations are responsible for providing supporting cost estimates as well as collecting actual cost data. BHI executes the D&D activities using Hanford Plant Force union workers, except for construction work, which falls under the requirements of the Davis-Bacon Act. Davis-Bacon construction-type activities are bid competitively and awarded to firms in accordance with established DOE FAR. BHI provides project management, engineering, compliance, regulating health and safety, quality assurance, and field support for the project.

In developing a list of potential partners for the Strategic Alliance, did DOE issue a CBD announcement for an expression of interest concerning the cocooning of C-Reactor?

The selection of the members of TAT was made on the basis of previous D&D experience and a desire to become a part of the C-Reactor LSDP. A CBD solicitation was used. CH₂MHiLL, International Technology, and Thermo Electron are part of the Environmental Restoration Contractor and they represent the corporate offices. AEA Technologies is a European firm and provides the TAT connections to D&D activities internationally. The D&D subgroups also include representation from local Indian tribes, stakeholders (primarily DOE via the Hanford Site Technology Coordination Group [STCG]), and regulators (Washington State Department of Ecology and the EPA). METC required that the U.S. Army Corps of Engineers be represented on the TAT, and Montgomery-Watson and Morrison Knudson are the Corps representatives.

How does DOE-RL make cost estimates for the D&D activities? How are the cost-sharing technology demonstrations done?

Delphinus Engineering provided a detailed cost estimate and man-rem exposure estimates during development of the CDR and later the 12-volume detailed D&D plan. Baseline technologies were used for this cost estimate. After technology demonstrations are awarded, the technology providers are required to submit actual cost data to the I/C in sufficient detail for comparison with the baseline costs. Cost data is collected and made a part of the Innovative Technology Summary Report (ITSR).

The C-Reactor project is not being carried out on a cost-sharing basis with the technology providers because it is felt that small business will not receive a "fair shake" if it is required to share the cost. EM-50 is funding the technology demonstrations. Thirty percent of the EM-50 funding (30 percent of the approximately $9 million EM-50 is providing over the life of the project) is targeted for allocation to private industry for technology demonstrations.

In developing the plans for the D&D of the C-Reactor, did DOE-RL rely on documentation associated with the D&D of the Shippingport Reactor?

Delphinus Engineering prepared the C-Reactor D&D plan. Personnel who had some involvement in the Shippingport Atomic Power Station D&D now work for Delphinus and were used during the preparation of the C-Reactor D&D plan. Bechtel personnel who were involved in the Three Mile Island Plant 2 cleanup and recovery effort also participated in the plan development and review.

How will DOE-RL prevent getting involved with an immature technology, that is, a technology which is not sufficiently developed to be tested as an LSDP?

The Technology Provider must then submit detailed procedures for review and approval. Immature technologies should be identified at this step.

Under which set of Environmental Safety and Health (ES&H) regulations (such as 10CFR835) do the activities at C-Reactor fall?

The C-Reactor Interim Safe Storage Project is being conducted by BHI under CERCLA requirements and DOE orders. The Hanford Environmental Remediation Disposal Facility (ERDF) is also permitted as a CERCLA facility, although it was designed to RCRA criteria. ERDF can accept up to 1.2 million cubic yards of waste. At present approximately 6,000 cubic yards per week are being sent to the ERDF. DOE EM-30 is not involved in the C-Reactor D&D, since any waste generated will be sent to ERDF.

Necessary and sufficient ES&H standards are not being used for the C-Reactor LSDP. All work is done in accordance with radiological regulations (i.e., 10CFR835), applicable DOE orders, and federal/state regulations.

Technology selection and evaluation:

- What is the baseline technology of the D&D project?
- What are the criteria for selection of LSDP over the baseline technology?

- What are the measurement criteria used to evaluate the LSDP in relation to the baseline technology?

The process for technology selection is shown in Figure 2. Hanford has identified six focus areas for which innovative technologies will be selected, used, and evaluated. The Hanford focus areas are characterization, decontamination, demolition, waste disposition, facility stabilization, and worker health and safety. It is important to note that BHI plans to evaluate the innovative technology (LSDP) in a direct side-by-side comparison with the baseline technology. For example, if the activity is decontaminating the surface of a wall, both the innovative technology and the baseline technology will be used to clean comparable size areas of the same wall.

Technologies that already have undergone demonstration at another DOE site will not be retested at C-Reactor unless there is a compelling reason to do so. For example, although the self-contained air-cooled respirator suits have been evaluated at CP-5, the demonstration did not utilize the suits in an actual reactor D&D situation. Since there is a need at C-Reactor for this protective equipment, it will be reevaluated. Four characterization technologies demonstrations are scheduled to start into the procurement cycle in mid-November 1996.

Is there a standardized format for documentation and reports?

Cost and performance data are collected and made a part of the ITSR. The ITSR guidelines help ensure that a uniform, standardized format is used for all LSDPs in the EM-50 program.

How are the I/C Teams selected? Are there established criteria?

The selection of I/C Teams members was not based on a competitive process. Members of the I/C Team are prohibited from bidding on large-scale demonstration projects.

How does DOE ensure that industry knows what technologies DOE needs for future D&D projects that can be demonstrated in an LSDP?

The Hanford LSDP program also uses the CBD, the Internet, the STCG, and Environmental Technology Partnerships to seek companies that might have innovative technologies to demonstrate. In seeking companies, the problem to be addressed is defined and one or more vendors are sought. Demonstration Packages are prepared with no specific Technology Provider in mind; only the problem to be addressed by the innovative technology is specified.

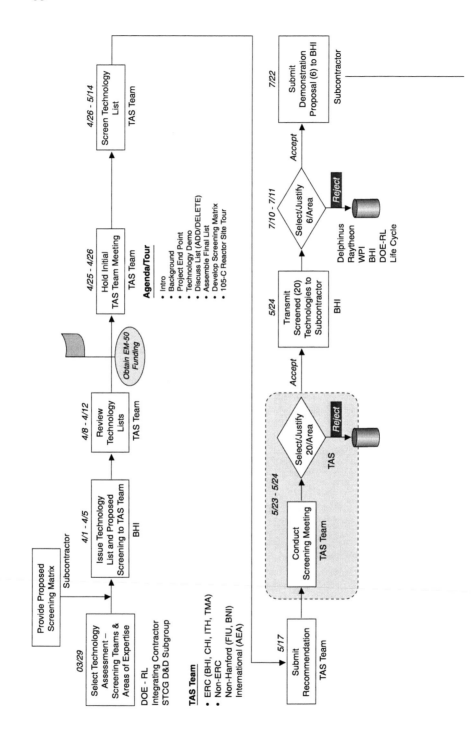

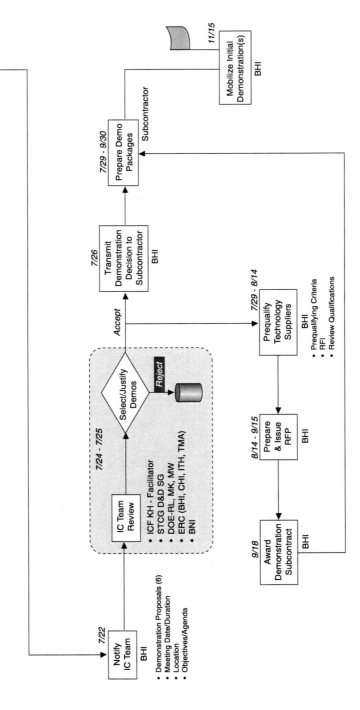

FIGURE 2 Technology selection process.

How were liability issues resolved?

BHI indicated that to date, they have not encountered any liability issues with the implementation of the C-Reactor LSDP. BHI is following the contractual requirements of its contract with DOE-RL by including appropriate contract clauses and requirements to the members of the TAT and other subcontractors. Product liability and intellectual property rights have not been barriers in setting up the C-Reactor LSDP.

How were intellectual property rights handled?

Intellectual property rights are retained by the Technology Provider.

Summary of the meetings held on March 13-14, 1997, at the Fernald Environmental Management Project on the subject of the DOE Large-Scale Technology Demonstration Program associated with the D&D of Plant 1

Trip Report by
Linda Wennerberg and Alfred Schneider for the
NRC Committee on Decontamination and Decommissioning
Submitted July 20, 1997

Site Host
Mark Peters   Fluor Daniels Fernald

Presenters
Mark Peters   Fluor Daniels Fernald
Brad Connley   Fluor Daniels Fernald
Dick Martineit   Fluor Daniels Fernald
Grace Ruesink   Fluor Daniels Fernald
Larry Stebbins   Fluor Daniels Fernald
Terry Borgman   Fluor Daniels Fernald
Steve Bossart   DOE-FETC, Morgantown, WV
Kevin Jones   U.S. Army Corps of Engineers
Fred Huff   U.S. Army Corps of Engineers

THE FERNALD LARGE-SCALE TECHNOLOGY DEMONSTRATION PROJECT

Introduction

The Fernald LSDP was one of the first three LSDPs selected by the DDFA of the DOE's OST. The direction of this project was through the FETC in Morgantown, West Virginia.

The Fernald Environmental Management Project (FEMP) is an inactive DOE facility near Cincinnati, Ohio, formerly used in the production of uranium and thorium metal and other compounds. It is a Superfund site currently undergoing environmental restoration. The contractor is Fluor Daniels Fernald (FDF).

The LSDP was to be integrated into the Plant 1 D&D Project, which was already in progress under a fixed-price contract awarded to B&W Nuclear Environmental Services, Inc. The total cost was estimated at $10.9 million and the project was to last 18 months.

Objectives

The completion state of the Plant 1 D&D Project was removal of all structures down to their concrete pads, with the transite panels palletized and wrapped and the structural steel segmented and stacked on the Plant 1 pad. Debris was to have been boxed for disposal on site, along with the transite and structural steel. Debris not meeting the FEMP waste acceptance criteria was to be sent to the Nevada Test Site (NTS).

The technologies selected for demonstration were to be judged for their values in:

a) Reducing the size of the labor force required (major cost factor);

b) Improving safety and reducing the probability for work-related injuries or accidents; and

c) Reducing the volume of waste requiring off-site disposal.

LSDP Initiation

FETC issued a Request for Letter Proposals on July 21, 1995, and a proposal was submitted by the DOE Fernald Office on August 17, 1995. The proposal was accepted in October 1995 for an estimated cost of $3.1 million. The period for implementing the LSDP was to be October 2, 1995, to July 20, 1997. The award made it mandatory that at least 40 percent of the funding go to technology vendors in the private sector.

LSDP Organization

Because of the existing organization for the Plant 1 D&D Project, the LSDP consisted of two parallel organizations as shown in Figure 1. Important organizational objectives were to avoid critical path interference and to secure the necessary support of the ongoing D&D project. An important feature of the organization was the formation of the Integrating Core Team which included the FDF Plant 1 Project Manager and the Plant 1 D&D Contractor. This team provided direction for the LSDP and approval of all the technology demonstrations.

Technology Selection

The LSDP developed a D&D "Needs Statement" specifically for the FEMP D&D tasks with inputs from all parties involved in D&D activities at the site. The technologies identified included:

- Pipe shearing equipment for remote removal of pipes and conduits;
- Personal protection equipment;
- Improved methods for power washing of structures;
- Improved concrete scabbling methods;
- Equipment for inspection and cleaning of process piping;
- Enhanced removal of insulation, transite panels, and concrete/masonry walls; and
- Instruments for in-situ characterization of surface contamination.

Detailed criteria were used during the evaluation process. Of the 184 technologies that were screened, 32 passed, 17 were proposed, and 12 were approved for demonstration. Subsequently, two approved demonstrations were canceled: a passive aerosol generator for the removal of loose surface contamination because of incompatibility of the wall coatings available, and a transite pulverizer/transfer system because of scheduling conflicts.

Results

The following results were reported:

- Vacuum removal of insulation—superior to the baseline method (manual removal and bagging) by lowering airborne contamination, waste volumes, number of bags, and man-hours required; removal rate was about 25 percent faster than the manual method.
- Washing of D&D debris—the steam/air spray vacuum technique (Kelly Decon System) was found unsuitable for small debris segments and did not offer any advantage over the baseline method (high-pressure water).

74

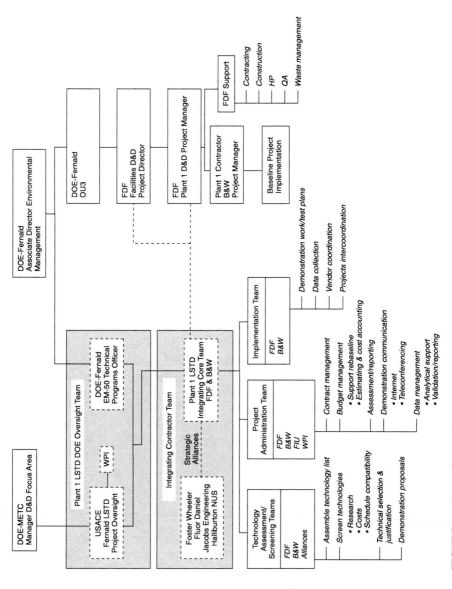

FIGURE 1 Plant 1 Large-Scale Demonstration Organizational Structure.

• Sponge cleaning of equipment—generally found adequate for cleaning contaminated process equipment, especially if it contained enriched uranium. System was found to have potential benefits. No baseline method was available for comparison.

• Raman spectroscopy—method for the radiological characterization of swipes, including isotopic information, was found to be unsuitable.

• Laser induced fluorescence—for rapid determination of uranium surface contamination. The principle was successfully demonstrated, but there is a need for additional development work.

• Passive aerosol generator—for the removal of surface contamination. Demonstration was canceled.

• Void filling with polyurethane foam—demonstration was successful, but the use of low-density cellular concrete was deemed preferable.

• Void filling with grout foam—for the specific application in which subsidence is to be avoided, this method and material was found to have benefits over the baseline method, which requires the segmenting of metal equipment having voids greater than 1 cubic foot.

• Cool suit for worker protection—demonstration postponed for seasonal reasons.

• Oxy-gasoline torch—significant advantages found over the baseline equipment (acetylene cutting torch).

• Pipe inspection—an inspection system using a miniature video camera showed cost benefits for this particular project by qualifying the removed process piping for on-site burial rather than shipment to the NTS.

• Transite pulverizer and transfer system—demonstration was canceled.

Dissemination of Information

Efforts were made by DOE and the site contractor to disseminate the demonstration results through reports, the Internet, D&D contractor briefings, presentations at conferences and topical meetings, and specific references in bids for D&D proposals.

Cost Benefits

An elaborate system was established for the development of cost estimates for the demonstrated technologies for comparison with the actual costs incurred when the baseline technologies were used. The U.S. Army Corps of Engineers was given the prime responsibility. Since many assumptions and corrections have to be made, the resulting uncertainties may either mask or exaggerate any real differences in the costs of these technologies. Unless there are very substantial differences in man hours, waste volumes, and equipment costs, which was not the case for any technologies of the Fernald LSDPs, purported benefits will be difficult to justify on a cost-saving basis.

Conclusions

1. The Fernald LSDP was well planned, the organizational structure met its objectives, and a dedicated and well-qualified staff performed efficiently.

2. The fundamental problem in selecting technologies deserving of an LSDP is shown by the fact that of 184 technologies screened, only 12 were found to conform with the acceptance criteria, and for the majority of the 10 technologies actually demonstrated, the results could have been predicted quite reliably without the elaborate procedures employed in the course of the demonstrations.

3. It will be difficult to accept these cost comparisons, since many assumptions and corrections had to be made during their development.

4. It is not known how effective the dissemination of the results was.

5. The cost effectiveness of the Fernald LSDP is probably low, considering that for an investment of $3-4 million the actual savings resulting from the use of the demonstrated technologies are likely to be small.

6. The Fernald LSDP showed that the usefulness of a particular method depends greatly on local conditions (types of equipment, nature and extent of contamination, applicable regulations, etc.).

7. The most obvious benefit of the Fernald LSDP was in showing that a disciplined search and review of existing technologies by contractors for specific applications can be beneficial and should be encouraged by the DOE.

F

Summary of the meeting and tour held on July 30, 1997, at the Argonne National Laboratory on the subject of the DOE Large-Scale Technology Demonstration Program associated with the D&D of the CP-5 Reactor

**Trip Report by
Joe Byrd and Milt Levensen for the
CEMT Subcommittee on D&D
Submitted August 12, 1997**

Attendees

Samit K. Bhattacharyya	Associate Director, Technology Development Division, Argonne National Laboratory
Tom Yule	Program Manager, Technology Development Division, Argonne National Laboratory
Robert W. Rose	Project Manager, CP-5, Technology Development Division, Argonne National Laboratory
Dennis Haley	D&D Coordinator, EM-50 Robotics Program, Oak Ridge National Laboratory.
Joe Byrd	CEMT Subcommittee on D&D
Milt Levensen	CEMT Subcommittee on D&D

Documentation received prior to and during the meeting
Copies of overheads from presentations from Bhattacharyya and Rose
Technology Demonstration Summary Sheets (13 total)
Previous Trip Report on Subcommittee Visit to CP-5
Reference paper on ROSIE (from FETC D&D Review Meeting)
Reference paper on Dual Arm System (from FETC D&D Review Meeting)

INTRODUCTION

A complete report on the CP-5 LSDP was presented in a June 3, 1996, trip report from Trish Baisden and Frank Crimi on their visit to Argonne on April 18, 1996. The report includes the selection process for the LSDPs, discussion of the Strategic Alliance, technologies selection process, and technologies reporting process. This trip report will not repeat that detailed background information on the CP-5 project; it will reflect only information received and observations made during this brief visit. The primary purposes of this visit were to follow up on accomplishments since the previous visit at the beginning of the project and to report on the progress made, particularly in the use of robotics.

MEETING SUMMARY

An overview of the Argonne site and its "Vision of the D&D Technology Center" was presented. D&D technology is now a "thrust area" of the ANL. ANL proposes to organize a D&D Technology Center and function as the leader in the area of D&D technologies. The Center is an internal ANL concept and does not relate to FETC or any current EM program. It is a concept of bringing together the various capabilities of ANL to bear on a single problem, independent of the source of the funds. The Center has completed three D&D projects and is making good progress in meeting its objectives in the CP-5 D&D project. The three completed projects were successful under a very limited objective suitable for internal use of the facility. None were completed to a "free release" state, and none of the wastes were "disposed of." However, the end-state objective met ANL's needs. Much discussion was held concerning follow-through on a D&D project through storage, disposal, and/or recycle of the materials. At the present time these "total D&D" issues have not been addressed fully.

Information was presented on the CP-5 project, including ANL's involvement as a DDFA LSDP. From the beginning, ANL has approached CP-5 as a D&D project, not just a demonstration of technologies. It did not limit the choice of demonstrated technologies to EM-50 constraints. A review of the Strategic Alliance organization and participation, the technology selection/rejection processes, and the reporting process was also presented. The selection committee evaluated 88 vendors and 55 innovative technologies. Although all documentation refers to "innovative" technologies, it was pointed out that many of the technologies chosen were proven, off-the-shelf systems using little or no innovative technology. This fact was acknowledged; apparently, "innovative" is used throughout the documentation, although it is recognized that it does not apply to all systems. Twenty-one technologies were selected for demonstration during the CP-5 D&D project. Five of these were robotic technologies. These five are discussed in the following section. Preliminary results of all demonstrated technologies were presented. Thirteen technology demonstrations have been completed and one-page Technology Demonstration Summary Sheets have been published. The nine remaining demonstrations have been scheduled. The Green Book process has bogged down in its implementation. No Green Books for CP-5 have been published. Significant accomplishments, lessons learned, and future plans were reviewed. The final hour of the visit was a tour at the CP-5 facility.

CURRENT ROBOTICS DEMONSTRATIONS AT CP-5

Two robotics systems are currently at CP-5: the Dual Arm System and the ROSIE Mobile Work Platform from RedZone Robotics, Inc., Pittsburgh, Pennsylvania.

1. Dual Arm System: This system is currently in use to remove the reactor liner. The dual arms are commercial systems that can be configured for a specific application. ORNL developed the control console and software. INEL designed the deployment platform for use at CP-5. Unfortunately, the first failure had been experienced the evening before our visit and the system was undergoing maintenance. However, over 1,000 hours of trouble-free operation had been reported prior to that time. The operation and performance of this system has been judged to be reasonably satisfactory. There has been a learning curve associated with its operation (as with any new system) but operators have accepted it as an effective, useful tool. Tryouts were held for operation of this system. Approximately 70 percent of the operators made the first cut, and about two weeks were spent in training before actually using the system to perform work. The system has limitations since it was not designed specifically with CP-5 in mind. The system was designed to use commercial cutting tools (circular saw, router, etc.). The cutting tools were designed for 3/8-inch steel, whereas 3/4-inch materials exist at CP-5. This has resulted in a more time-consuming operation than originally estimated. Currently, the CP-5 is slightly behind schedule.

2. ROSIE Mobile Platform: This system is on the floor at CP-5 but not currently in use. This is a mobile system containing a high-weight-capacity, long-reach boom. It has been demonstrated and operated by ANL operators. As with the dual arms, operators were given a try-out test. Approximately 50 percent of the operators made the cut for this system. It is perceived to be more complex in operation than the Dual Arm System. This would be expected in a mobile system that introduces another degree of freedom. The CP-5 project was not a good demonstration theater for ROSIE as currently configured. This system was developed by RedZone Robotics and funded through the FETC industry program. The original proposal by RedZone included a dexterous arm at the end of ROSIE's boom. This was a requirement for typical dismantling tasks. The proposal funding was cut to eliminate the dexterous arm. When ROSIE arrived at CP-5 without the arm, an arm was purchased to be used with ROSIE. However, the procured arm at ANL is not compatible with the hydraulic fluid used in

ROSIE. Since there is no dexterous arm with tools, the system has very little use at CP-5. The work area around the reactor is too small for this large mobile system. It is perceived that with the proper tooling this could be a useful system for a variety of D&D operations.

Three other robotic technology demonstrations have been completed:

• Mobile Automated Characterization System (MACS): MACS was not demonstrated as part of the Strategic Alliance LSDP contract. It was an effort with ORNL during July 1996 prior to the beginning of the official LSDP. CP-5 was a poor choice as a demonstration site for this system. It is a mobile robot designed to autonomously characterize large floor areas such as those at Oak Ridge K-25. A small work area around the CP-5 reactor was used as the test bed. It performed well in that area, but could not be adapted to the circular walls, etc. It was too large to survey the desired areas around CP-5. However, it is perceived to have high potential usage in the environment for which it was designed.

• Pipe Crawler Radiological Surveying System: Pipe Crawler, a tele-operated system developed by Radiological Service, Inc., as part of a turnkey pipe inspection, decontamination, and survey service, was demonstrated in December 1996. Surveys were conducted primarily in the rod storage holes, along with portions of a pair of 12-inch vent lines servicing the reactor area. Several rod storage holes each of 5-, 6-, and 12-inch diameter and 10- to 17-foot depth were surveyed along with 40 feet of the combined vent lines. Surveys were preceded by video inspections but without the benefit of pipe pre-decontamination. These demonstrations are considered very successful with potential wide-spread usage for a variety of D&D projects. This is a commercial system available from a service provider.

• Pipe Explorer System: Pipe Explorer, developed by Science & Engineering Associates, Inc., is a teleoperated method for transporting characterization tools into pipes and ducts. It deploys a plastic membrane to provide a clean conduit for the sensors to travel through. This July 1996 demonstration was involved in alpha surveys of the internal area of three 5-inch pipeline holes in the CP-5 rod storage area. In addition, a video inspection and a beta/gamma survey of a 3-inch drain line in the CP-5 yard area were completed. These demonstrations are considered very successful with potential widespread usage for a variety of D&D projects. This is a commercial product.

GENERAL COMMENTS

• ANL has viewed the CP-5 project primarily as an ANL D&D Project and secondarily as the Focus Area LSDP. They had formalized plans and schedules for the CP-5 project prior to being selected as an LSDP site.

• The demonstrations have been beneficial in providing mechanisms and procedures for contract vendors to get onto nuclear sites and have provided them a better understanding of working at a nuclear facility.

- Operation of the Dual Arm System and ROSIE has provided information on the selection process and the training required for using on-site operators with new robotic equipment.

- The technologies selected for demonstration did not have detailed test plans prior to arrival at CP-5.

- DOE determined the makeup of the Strategic Alliance.

G

Biographical Sketches of the Committee

MYERS, PETER B. The D&D Committee chairman, Dr. Peter Myers, served in World War II with the U.S. Navy before completing his doctorate in nuclear physics as a Rhodes Scholar at Oxford University. He retired in 1993 after 14 years as director of the National Research Council's Board on Radioactive Waste Management. Dr. Myers has held management positions with responsibility for research and development of advanced solid-state technology at Bell Telephone Laboratories, Motorola, the Martin Company, Bunker Ramo, and Magnavox. Dr. Myers is a fellow of the Institute of Electrical and Electronic Engineers and the American Association for the Advancement of Science. He was a founding member of the Institute of Management Sciences and is a member of the American Nuclear Society, the American Physical Society, and the Materials Research Society.

BAISDEN, PATRICIA ANN Dr. Baisden is a nuclear chemist and her areas of expertise include the chemical and nuclear properties of the actinide elements, radiochemistry, analytical sciences and materials characterization. After receiving her Ph.D. from Florida State University in 1975, she joined the Lawrence Livermore National Laboratory where she currently is the Division Leader—Analytical Sciences Division and Materials Program Leader—Laser Programs within the Chemistry and Materials Science Directorate. Dr. Baisden has served on several committees of the National Research Council as well as the National Science Foundation's Nuclear Science Advisory Committee. She also was chair of the American Chemical Society's

Division of Nuclear Chemistry and Technology. Dr. Baisden currently is an editor of the international journal, *Radiochimica Acta.*

BURSTEIN, SOL Dr. Burstein is a registered professional engineer and currently does consulting work mainly in the areas of nuclear and mechanical engineering management. He retired as vice chairman of the board of directors of Wisconsin Energy Corporation and spent 21 years with Wisconsin Electric Power Company. Dr. Burstein is the recipient of numerous awards and honors, including distinguished service awards from the University of Wisconsin, Marquette University, and the Engineering and Scientific Societies of Milwaukee. He has served on numerous industry and government advisory committees, including the Nuclear Safety Research Review Committee of the U.S. Nuclear Regulatory Commission, several committees of the National Research Council, and the Board on Radioactive Waste Management. Dr. Burstein is a member of the National Academy of Engineering. He received his B.S.M.E. from Northeastern University and an honorary D.Sc. from the University of Wisconsin, Milwaukee.

BYRD, JOSEPH S. Mr. Byrd is a registered professional engineer and currently is an expert robotics consultant for the Operating Engineers National Hazmat Program (OENHP), having retired in 1997 as Distinguished Professor Emeritus of Electrical and Computer Engineering, University of South Carolina (USC). He is a Director of Cybermotion, Inc., of Roanoke, Virginia. He has 35 years of experience in the areas of computers and robotics, including teaching, research, design, development, and management. Prior to his academic career, he organized and managed the Robotics Technology Group at the Savannah River Laboratory operated by E. I. Du Pont De Nemours for the Department of Energy. He has numerous publications and presentations and has served as general chair and session chairs for major robotics conferences. He has served on government advisory committees and several NRC committees. His memberships include the American Nuclear Society, the Institute of Electrical and Electronic Engineers, Eta Kappa Nu, and Tau Beta Pi. His honors and awards include several professional and teaching awards and the American Nuclear Society Ray Goertz Award for significant achievements in robotics. He received his B.S.E.E. from Clemson University and his M.S.E.E. from the University of South Carolina.

CLEMENS, BRUCE W. Dr. Clemens recently joined the staff of James Madison University in Harrisonburg, Virginia, as a professor in the College of Integrated Science and Technology. Formerly, he was a member of the Tennessee Energy, Environmental, and Resources Center at the University of Tennessee, Knoxville. Dr. Clemens has managed multi-million dollar environmental programs around the world. While with the Environmental Protection Agency, Dr. Clemens was responsible for the first guidance on Remedial Investigations and Feasibility Suites. He has acted as a consultant for federal, state, and international organizations. He has served on various national and international task forces on subjects including hazardous-waste management, disaster relief assistance, and public water supplies and has served as a technical expert in court cases. He received his B.S. in civil and environmental engineering at Cornell, his M.P.A. in economics at Harvard, and his Ph.D. in strategic management at the University of Tennessee.

CRIMI, FRANK, P. Mr. Crimi recently retired as vice president of Lockheed Martin Advanced Environmental Systems Company. He joined Lockheed in 1992 after completing 34 years with the General Electric Company (GE). Mr. Crimi has over 38 years of experience in the design, operations, and maintenance of nuclear power plants with special emphasis in decontamination and decommissioning of nuclear facilities. He held a number of key management positions at the DOE Knolls Atomic Power Laboratory (operated by GE). His experience includes the management of large, complex programs in the nuclear industry, including construction, operation, and maintenance of naval nuclear reactor plants, and manager of the decommissioning services for GE. He was the GE program manager for decommissioning DOE's Shippingport Atomic Power Station and recently chaired the Long Island Power Authority's Independent Review Panel during the decommissioning of the Shoreham Nuclear Power Station. Mr. Crimi was a member of Public Service of Colorado's Management Oversight Committee during the decommissioning of the Fort Saint Vrain Nuclear Generating Station. He is currently a member of the Nuclear Safety Assessment Board, which provides oversight during the decommissioning of the Connecticut Yankee Power Company's Haddam Neck Nuclear Plant. Mr. Crimi was a member of the NRC's Committee on Decontamination and Decommissioning of DOE's Uranium Enrichment Facilities. He currently also serves as a member of the Committee to Review DOE/OST's Peer Review Process. He holds a B.S. in mechanical engineering from Ohio University and did graduate studies in mechanical engineering at Union College, Schenectady, New York.

LEVENSON, MILTON Since his retirement as vice president of Bechtel International, Mr. Levenson has done consulting work in chemical engineering with an emphasis on reactor safety, water reactor technology, fuel cycle technology, and breeder reactor development. He was the first director of nuclear power for the Electric Power Research Institute. He received the Robert E. Wilson Award of the American Institute of Chemical Engineers. Mr. Levenson is a fellow of the American Institute of Chemical Engineers and the American Nuclear Society and is a member of the National Academy of Engineering. He received his B.Ch.E. from the University of Minnesota.

SANDBERG, RAY O. Until his recent retirement, Mr. Sandberg was a project manager with Bechtel National, Inc., where he managed its design and cost-estimating team. He has an M.S. in chemical engineering and an M.B.A. He worked on the construction of a production reactor and in the major planning of the design of a heavy water reactor. Mr. Sandberg was responsible for post-accident planning for the recovery of Three Mile Island Unit 2, including testing of proposed decontamination techniques.

SCHNEIDER, ALFRED Dr. Schneider is currently an Emeritus Professor of Nuclear Engineering at Georgia Institute of Technology. Between 1991 and 1996 he served as Visiting Professor and Research Affiliate at the Massachusetts Institute of Technology. Prior to these positions, Dr. Schneider was Director of Nuclear Technology of Allied General Nuclear Services in South Carolina. Dr. Schneider has 47 years of experience in materials and process research and development, and operations and technical management of chemical and nuclear projects. He has consulted to the U.S. government, the states of New York and Georgia, industry, and nonprofit organizations in the areas of radioactive waste management and the nuclear fuel cycle. He is a member of the American Chemical Society, American Institute of Chemical Engineers, American Nuclear Society, American Association for Advancement of Science, Sigma Xi, Tau Beta Pi, and Phi Lambda Upsilon. He has served on three NRC committees and numerous advisory and oversight boards. Dr. Schneider earned his Ph.D. from the Polytechnic University of New York.

WENNERBERG, LINDA Dr. Wennerberg earned a Ph.D. in environmental law at Michigan State University in 1984. A private consultant with Environmental Business Systems, Dr. Wennerberg has 18 years experience in reviewing and developing environmental and economic policies with technical application; performance review of a federal toxic program; implementation of radioactive and hazardous waste management programs for state agencies; development of the draft siting criteria process for low-level radioactive waste disposal; determination of environmental enforcement priorities for oil and natural gas production; and identification of pollution prevention opportunities in manufacturing operations.

APPENDIX

H

Acronyms and Abbreviations

AME	Assistant Manager for Environmental Restoration
ANL	Argonne National Laboratory
ASTD	Accelerated Site Technology Deployment
BHI	Bechtel Hanford, Inc.
BOA	Basic Ordering Agreement
BOD	Board of Directors
CBD	*Commerce Business Daily*
CDR	Conceptual Design Report
CEMT	National Research Council's Committee on Environmental Management Technologies
CERCLA	Comprehensive Environmental Response, Compensation, and Liability Act (or Superfund)
ComEd	Commonwealth Edison
CP-5	Chicago Pile-5
CY	calendar year
D&D	decontamination and decommissioning
DDFA	Decontamination and Decommissioning Focus Area
DOE	U.S. Department of Energy
DOE-CH	U.S. Department of Energy Chicago Field Office
DOE-HQ	U.S. Department of Energy Headquarters
DOE-RL	U.S. Department of Energy Richland, Washington, Field Office
EBWR	Experimental Boiling Water Reactor
EIS	Environmental Impact Statement

EM	U.S. Department of Energy, Office of Environmental Management
EMAB	Environmental Management Advisory Board
EOI	expression of interest
EPA	U.S. Environmental Protection Agency
ERDF	Environmental Remediation Disposal Facility
ES&H	Environmental Safety and Health
FAR	Federal Acquisitions Record
FDF	Fluor Daniels Fernald
FEMP	Fernald Environmental Management Project
FETC	Federal Energy Technology Center (formerly METC)
FIU	Florida International University
FY	fiscal year
GAO	General Accounting Office
I/C	Integration Contractor
IFR	Integrated Fast Reactor
INEEL	Idaho National Environmental and Engineering Laboratory (formerly Idaho National Engineering Laboratory [INEL])
ITSR	Innovative Technology Summary Report (also known as the "Green Books")
LANL	Los Alamos National Laboratory
LLNL	Lawrence Livermore National Laboratory
LSDDP	Large-Scale Demonstration and Deployment Program (the LSDP was renamed in 1998)
LSDP	Large-Scale Technology Demonstration Program
M&S	maintenance and surveillance
MACS	Mobile Automated Characterization System
METC	Morgantown Energy Technology Center (now FETC)
N&SS	Necessary and Sufficient Standards
NES	Nuclear Engineering Services
NRC	National Research Council
NTS	Nevada Test Site
ORNL	Oak Ridge National Laboratory
OST	Office of Science and Technology (formerly the Office of Technology Development)
OTD	Office of Technology Development (now the Office of Science and Technology)
R&D	research and development
RCRA	Resource Conservation and Recovery Act
ROD	Record of Decision
SAR	Safety Analysis Report
SARA	Superfund Amendments and Reauthorization Act
SRS	Savannah River Site

STCG	Site Technology Coordination Group
Superfund	Comprehensive Environmental Response, Compensation, and Liability Act (CERCLA)
TAT	Technology Alliance Team
TDI	Technology Development Initiative
TP	technology provider
TRU	transuranic
TST	Technology Selection Team
TTP	Technology Test Plan
WBS	work breakdown structure